Walter Keeping

The Fossils and Palæontological Affinities of the Neocomian

Deposits

of Upware and Brickhill (Cambridgeshire and Bedfordshire)

Walter Keeping

The Fossils and Palæontological Affinities of the Neocomian Deposits
of Upware and Brickhill (Cambridgeshire and Bedfordshire)

ISBN/EAN: 9783337012830

Printed in Europe, USA, Canada, Australia, Japan

Cover: Foto ©berggeist007 / pixelio.de

More available books at **www.hansebooks.com**

THE FOSSILS

AND

PALÆONTOLOGICAL AFFINITIES

OF THE

NEOCOMIAN DEPOSITS

OF

UPWARE AND BRICKHILL.

London: C. J. CLAY, M.A. & SON,
CAMBRIDGE UNIVERSITY PRESS WAREHOUSE,
17, Paternoster Row.

CAMBRIDGE: DEIGHTON, BELL, AND CO.
LEIPZIG: F. A. BROCKHAUS.

THE FOSSILS

AND

PALÆONTOLOGICAL AFFINITIES

OF THE

NEOCOMIAN DEPOSITS

OF

UPWARE AND BRICKHILL

(CAMBRIDGESHIRE AND BEDFORDSHIRE)

WITH EIGHT PLATES

BEING

THE SEDGWICK PRIZE ESSAY FOR THE YEAR 1879

BY

WALTER KEEPING, M.A. F.G.S.

" Speak to the earth, and it shall teach thee :
And the fishes of the sea shall declare unto thee."
JOB xii. 8.

CAMBRIDGE:
AT THE UNIVERSITY PRESS.
1883

Cambridge:

PRINTED BY C. J. CLAY, M.A. AND SON,
AT THE UNIVERSITY PRESS.

PREFACE.

THE study of the Neocomian faunas of Cambridgeshire and Bedfordshire has occupied much of my time for some years past. I have had the advantage of watching the course of the 'coprolite' workings from their beginning in 1866 up to the present time, and have constantly been familiar with the large collections that have been made by the Woodwardian Museum, by Mr J. F. Walker of Sidney Sussex College, Mr E. Earwaker of Merton College, Oxford, and other Geologists.

In the comparison of the fossils with known species it has always been my endeavour to see the original or, failing that, some typical specimen, and to trust as little as possible to bare figures and descriptions. In this work I am much indebted to a number of Geologists for their kind assistance, for the loan of specimens, &c., amongst whom I must particularly mention Mr J. F. Walker, M.A., F.G.S., Mr C. J. A. Meÿer, F.G.S., Mr J. J. Harris Teall, M.A., F.G.S., Mr E. C. Davey, F.G.S. of Wantage, and Mr T. Davidson, F.R.S. of Brighton. In continuing the same work of comparison and identification through Holland and Germany I had the advantage of Professor Judd's Paper[1] as an admirable guide, and I am much beholden to Professors Marten of Leiden, R. Lepsius of Darmstadt, Ulrich of Hanover, and Geinitz of Dresden, and other German Geologists for their kindness and valuable help.

---

[1] "On the Neocomian strata of Yorkshire and Lincolnshire, with notes on their relations to the beds of the same age throughout Northern Europe." *Quart. Journ. Geol. Soc.* Vol. xxvi. p. 326.

A considerable number of the species from Upware and Brick-hill prove to be as yet undescribed, a fact which was to be expected in so isolated and peculiar a deposit—so thoroughly 'episodal' as Mr Blake would express it—as that of Upware and Brickhill.

In working with the already known species the reference to the original figure has always been made; and I have also given, when possible, references to some few other good figures and descriptions such as may be most useful or accessible to working Geologists; but no attempt is made to work out the complete synonymy of each species. Such a work can indeed rarely be quite satisfactory, depending, as it must do, to so great an extent upon the comparison of figures and descriptions only.

In the nomenclature it will be found that I have in several cases adopted names and used them as of varietal value and not as distinct species—e.g., *Ostrea frons*, Park, var. *macroptera*, Sowerby, for it appears to me that such a trinomial system is a growing necessity in many of the larger generic groups, both in recent and fossil organisms.

Amongst the matters of more general interest worked out in these pages will be found :—

(1) The close palæontological relationship of the Ironsand and Phosphatic series as found at Upware, Potton, Brickhill, and Farringdon.

(2) The *special* character of the native forms of life in our Lower Greensand Phosphatic beds ;—their richness in Brachiopods, Polyzoa, and Sponges.

(3) The influence of different physical conditions upon the characters of the faunas as illustrated by the Upware and Potton fossils (p. 48).

(4) The presence at Upware of a little batch of species which flourished long afterwards in the Upper Chalk period in the neighbourhood of Dresden (pp. 20, 119),

(5) The curious resemblances of the Upware group of oysters to the well-known Jurassic species *O. dilatata, deltoidea, nana* and *gregaria.*

(6) The profusion of Brachiopod shells, both species and specimens at Brickhill and Upware and the graduation of the various types (species) into one another (p. 22).

(7) The similarity of the Upware and Brickhill fossils to those of the Neocomian beds of the Brunswick area at Shöppenstedt and Berklingen (p. 73).

(8) The existence of a large 'derived' fauna in the coprolite beds, these being to a great extent much worn and otherwise mutilated remains of shells, &c., washed out of the rocks of the old coast lines, of Neocomian to Oxfordian age.

(9) The very general—almost invariable—phosphatization of these remains.

(10) The great similarity of the 'derived' Neocomian phosphatic nodules over wide areas.

(11) The occurrence of a 'derived' Neocomian fauna in beds of very nearly the same age, and the evidence of the rapidity of their fossilization, exhumation and redeposition.

(12) The evidence that the Vertebrate remains of Upware are, in great part, truly Neocomian species, native to the deposit in which they are found; while others are *derived.*

(13) The curious difficulty in determining the age of some of the Fishes' teeth; and the probable identity of form of some of the palatal teeth of Jurassic and Neocomian species; and

(14) The importance of distinguishing the Downham Market Phosphate Bed from the Ironsand and Phosphatic series as belonging to a separate Physical Group (pp. 11, 54).

My general conclusions as to the age of the Ironsand and Phosphatic series are in near accordance with the opinions of MM. Walker, Teall, Meÿer and Barrois, all of whom have placed the Upware bed in the Upper Neocomian or Aptien series (Lower Greensand)[1]. There is however some difference of detail between my results and those of Mr Teall. I refer them to a somewhat lower horizon of the Lower Greensand than does Mr Teall, believ- ‚ ing them to be of the age of Sandgate and Hythe beds with the lower part of the Folkestone series (Meÿer). This distinction becomes important when we take into consideration the views of some Geologists as to the relation of these beds to the superposed series; for the beds which are known as Upper Folkestone, the *mammillaris* zone, may well be taken as the basement bed of the gault just as we may regard the Cambridge Greensand as the basement bed of the chalk; and such I consider is the true aspect of the Downham Phosphate bed.

But our Ironsand and Phosphatic series (of Upware, Farringdon, &c.) I regard as truly belonging to the lower Group (Upper Neocomian or Aptien) which is very generally separated by a distinct break from the overlying Gault series—namely by the 'Great Unconformity' of Mr Judd as seen in Yorkshire, Lincolnshire, Hunstanton, Upware and Farringdon.

[1] Professor H. G. Seeley regards them as much older.

CONTENTS.

PART II.

SPECIAL PALÆONTOLOGY.

I. VERTEBRATA.

II. INVERTEBRATA.

TABLE OF CRETACEOUS STRATA.

	South of England (Vectian type)	Cambridgeshire, Bedfordshire, and elsewhere	Lincolnshire (Tealby type)	Yorkshire (Speetonian)
D. CHALK	D. CHALK { Upper chalk, with flints / Middle chalk / Lower chalk, without flints } *Chalk Marl* / *Chloritic Marl*	D. CHALK { Upper chalk, with flints / Middle chalk / Lower chalk, without flints } *Chalk Marl* / *Cambridge Greensand*	D. CHALK { Upper / Middle, with flints / Lower, without flints }	D. CHALK { Upper chalk, without flints / Lower chalk, with flints } *Grey chalk*
C. UPPER GREENSAND	C. UPPER GREENSAND *Upper part of Folkestone series with Mammillaris zone*	C. absent	C (with part of B). RED CHALK *Coprolite Bed*	C—B. RED CHALK
B. GAULT	B. GAULT	B. GAULT { *Downham Market and Phosphatic deposit* }		
A. NEOCOMIAN	II. Lower Greensand { iii. Folkestone series / ii. Sandgate and Hythe series / i. Atherfield clay } ; I. Weald clay and Hastings sand	A. II, ii and part of iii. *The Iron sand and phosphatic series of Upware, Potton, Brickhill, Farringdon, Hunstanton, &c.* ; Absent.	A. III, Upper Neocomian Sands / A. II, Middle Neocomian or Tealby Series (Limestones, Ironstones, and clay) / A. I, Lower Neocomian Sands *Coprolite Bed*	A. III; Upper Neocomian clays (Judd) / A. II, Middle Neocomian clays / A. I, Lower Neocomian clay *Coprolite Bed*

CHAPTER I.

GENERAL ACCOUNT OF THE DEPOSITS.

I. HISTORICAL.

BEFORE the years 1866—1867 little was known of the forms of life in the Lower Greensand of East-central England. But at that time the "thin and anomalous but highly interesting deposits of Upware and Potton"[1] were found to contain certain gravelly and conglomeratic beds of Phosphatic Nodules in such quantity as to be of commercial value. The work of extracting such nodules and their manufacture into agricultural manure was already an industry of considerable extent in the pebble bed above the gault (Cambridge Greensand) in the immediate vicinity of Cambridge, and the same work was soon established, and energetically carried on for some three or four years in the newly found *Lower* Greensand 'coprolite' beds beneath the gault at Upware.

These workings furnished to Palæontologists unusually good opportunities for the collecting of fossils, in a district which had hitherto been conspicuous for its barrenness in organic remains. Large collections were therefore gathered together, especially by the Woodwardian Museum, by Mr J. F. Walker of Sidney Sussex College and Mr E. P. Earwaker of St John's College. Many of the specimens were found by the collectors themselves upon the pre-

---

[1] Upware is a small village on the river Cam, about 10 miles from Cambridge and 5 miles from Waterbeach. It is in the parish of Wicken, and the two names Upware and Wicken have both been used, synonymously, in descriptions of this bed. There is a convenient inn, with the sign "Five miles from anywhere. No hurry!"

pared heaps of Phosphatic nodules or rejected rubbish, but the great majority were saved by the workpeople whose attention was directed to the " curiosities," and their interest maintained by the usual stimulus.

After the year 1868, the work gradually slackened and was soon almost entirely discontinued; but some new workings having been opened near Upware last year the sections are now (Aug. 1879) again well exposed to view.

The Potton deposit did not prove to be very rich in native forms of life but contained much the same kinds of fossil remains as are commonly found upon our present sea beaches (Folkestone for example), that is to say, a large quantity of pebbles and rolled fossils that have been washed down out of the neighbouring cliffs; a few shells which lived upon the spot; and the bones, mostly water-worn, of the vertebrated animals which lived upon the neighbouring lands.

But at Upware, owing perhaps to quieter waters and also to the calcareous nature of the shore line, or reef, of coral rag, a much richer aquatic fauna flourished, the study of whose remains has furnished the principal part of the work of the following pages.

In the year 1875, some new phosphatic workings, in beds of the same general character as at Upware, were opened out near the village of Little Brickhill, about 2½ miles east of Bletchley[1]. These workings were described by me in a paper published in the *Geological Magazine* of that year (p. 372), and I have nothing to add to the stratigraphical details of that section; our knowledge of its fauna has however largely increased since then, so that the Brickhill bed is now second only to Upware and Farringdon in its organic richness. Some of the rarer fossils from this place were got from certain calcareous nodular masses which occur scattered through the sands, but most of them were obtained in the usual way here, namely from the workmen, who find and preserve the fossils in the course of their employment of digging and sifting out the 'coprolites' from their sandy matrix; also from the women and children who are engaged as 'pickers' to weed out all the stony rubbish (quartz, Lydian stone, chert, &c.) from the true phosphatic pebbles before it is ready for the market.

[1] The nodules were found here first by Mr J. J. Harris Teall, of St John's College

II. Account of the Section at Upware.

The Upware deposit was first described by Mr J. F. Walker, M.A., in the *Geological Magazine* (Vol. IV., p. 310), and another account has since been published by the same author, in a foot-note to Dr Lycett's description of *Trigonia Upwarensis*, Lycett (Monog. Trigoniae Pub. Palaeontographical Society, Vol. XXIX., p. 145). An account of the overlying gault accompanied by a woodcut section was published by my father, Mr H. Keeping, in the *Geological Magazine*, Vol. V., 1868, and the whole subject has received special attention from Prof. Bonney, in his *Cambridge-shire Geology*, 1875, p. 22, and Appendix I. Also this bed to-gether with the Potton rock formed the subject of the Sedgwick Prize Essay for 1873, by my friend Mr J. J. Harris Teall of St John's College (1875 Pub.). To these various publications I must refer for all special particulars of the stratigraphical characters of the deposit; it will suffice here to indicate the general characters of a typical section, so as to shew the general nature of the deposit, and the relation of the beds to one another, and to the adjacent formations.

The appended section is in the main similar to one published by my father in 1868 (*Geological Magazine*).

(A) Lowermost we have the Coral Rag of Upware, a highly coralline rock as seen in the line of our section, but in some other places it is more compact in texture and more oolitic and arenace-ous in composition. In a large limestone quarry near by this rock is seen dipping to the N.W. at about 4° (recorded by Prof. Bonney).

(B) Upon this rests (conformably, as there is every reason to believe) the Kimmeridge Clay in its usual character. But for some depth around the present outcrop of the Coral Rag this latter rock has been bared of its covering of Kimmeridge Clay so that the phosphatic bed of the Lower Greensand comes to overlap on to the coral rag. The destructive work of the removal of the Kimmeridge Clay went on during the earlier times of the formation of the Lower Greensand, and one of its results was the production of a curious deposit composed of irregular broken fragments of

1—2

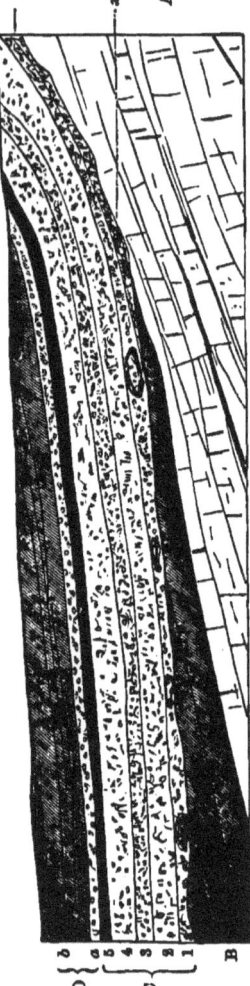

THE UPWARE SECTION.

(D) Gault.

　　b. Unfossiliferous gault—7 feet seen.

　　a. Phosphatic nodule bed, rich in fossils, about 5 inches.

Unconformity.

(C) Lower Greensand (= upper Neocomian or Aptien), about 12 feet.

　5. Clay-bed—about 1 foot thick (has been referred to the gault).

　4. Upper sand bed, or yellow 'silt.'

　3. Upper nodule bed.

　2. Lower sand bed, or yellowish 'silt.'

　1. Lower nodule bed and conglomerate.

　x = Junction bed of Lower Greensand, composed of Coral Rag fragments in a paste of Kimmeridge Clay.

Unconformity.

(B) Kimmeridge Clay.

(A) Coral Rag.

coral rag mixed up with the clayey material of the Kimmeridge Clay in such a manner that it "actually presents the appearance of Boulder Drift." This deposit is of limited extent, in the line of junction at the base of the Lower Greensand only.

(C) Then we come to the first definite bed of the Upware Neocomian—bed C. 1, of section—a pebbly sand and rocky conglomerate bed about one foot thick. The pebbles are mostly phosphate of lime, together with chert and Lydian stones, vein quartz, quartzite, jasper (less common) and coral rag (still rarer). The quartzites are of very various grain and colour, clear to opaque white, pink, yellow, and sometimes very dark; fragments of ironstone and iron sandstone are abundant, and some fine grained pink sandstones are occasionally found. All these rocks are very fragmentary and are more or less worn, some being quite rounded into pebbles, but the majority by far are still angular or subangular in their contours. These different stones, varying in size, mostly between the measurements of peas, beans, and walnuts are mingled together without order in a matrix of sand, with occasional oolitic grains of coral rag origin. Here and there these materials are cemented together by carbonate of lime to form irregular patches and nodules of conglomerate, in which many fossils were found. Each pebble in the hard conglomerate is enveloped in a filmy case of carbonate of lime which has formed around it in the matrix.

Mixed up with these pebbles are the shells of Mollusca, which, strange to say, are not at all worn but excellently well preserved, although some of them are of a very delicate nature ; and from this bed many of the best fossils were obtained, especially the indigenous Lamellibranchs and Gasteropoda. The richness of this bed in fossils, its hard conglomeratic character, and the excellent preservation of its shells are probably all connected with its contiguity with the coral rag.

(C. 2) Above the 'lower phosphate bed' just described comes a bed of loose sand, or 'silt,' red and yellow coloured, and composed for the most part of grains of quartz, Lydian stone, chert, and ironstone. Some of the quartz grains are blue and amethystine. Near the junction lines some irregular masses and nodules of irony and slightly phosphatic sand rock are found, which we again refer to in the sequel.

(C. 3) Next we come to the upper nodule bed which differs

from the lower nodule seam in the slightly paler colour of its phosphatic nodules, and in its scanty supply of carbonate of lime, so that this seam remains constant as a loose pebble bed.

The pebbles are the same as in the lower nodule bed. Amongst them the cherts and Lydian stones are particularly interesting. These vary greatly in colour, dark green to nearly black being most abundant; others are of lighter colours of yellowish and chocolate hues. Frequently they are found with bands of colour, or with projecting rings of more durable material. Many of the chert pebbles are, to the naked eye at least, precisely like chalk flints, both black and pale types being common[1].

It was in this bed that the workings for coprolites were principally carried on, and from it were obtained a large proportion of our Brachiopod fossils.

(C. 4)   The upper sand (C. 4) is similar to the lower seam (C. 2) between the two nodule beds, already described.

(C. 5)   The bed of clay next in order has been referred to the gault, but it is probably the representative of a bed of sandy clay or clayey sand belonging to the Lower Greensand which was seen in many of the Upware sections (e.g. the bed numbered (b. 4) in the Spinney Abbey section) as was suggested by Prof. Bonney in 1875 (Cambridgeshire Geology, p. 64). This is the more likely because phosphatic pebble beds (see bed Da in woodcut) occupy characteristically the lines of chronological breaks, especially throughout the cretaceous period.

[1] These chert and Lydian stone pebbles are very widely distributed in Neocomian strata. They are found at Upware, Potton, Brickhill, and Farringdon, and in the Bargate Pebble Beds, and they occur in precisely similar condition in the Hils conglomerat of Schoeppenstedt in Brunswick. They are also abundant in the Portlandian pebble bed of Bourton, near Swindon. Larger pebbles also occur, mostly of quartz and quartzite. Some of the pebbles, especially those from Potton, have yielded fossils, including shells, crinoids, &c. of carboniferous age from the chert; Jurassic shells and echinoderms from other kinds of chert, and a number of brachiopods (orthis, spirifera, &c.) from a pale shaly slate, apparently of age of the middle or Upper Bala group of Sedgwick. The details of these pebbles are given in a paper read before the Cambridge Philosophical Society, May 3rd, 1880, and published in the Geological Magazine, September, 1880.

The majority of these pebbles I regard as having been derived from an ancient Palaeozoic barrier axis which in the preceding, or Lower Neocomian period, separated the Northern from the Southern Neocomian seas in Europe, but which was in the time of the deposition of the Iron Sand series suffering rapid denudation and destruction.

Numerous variations from the above type of sections were found in the different workings around Upware. There may be three distinct phosphatic nodule beds and apparently all these may in places unite into one (Walker). The lower bed may be a simple pebble bed instead of being hardened here and there into a conglomerate, and it may be separated from the junction line of the jurassic rocks by the interposition of a bed of sand.

For the further illustration of the Upware deposit, under a somewhat different aspect, the new working recently opened out at Spinney Abbey is worthy of being described, for the sections in this border-district of the Upware area have never yet been recorded.

I observed this section in August, 1879, in company with Mr E. B. Tawney, M.A., F.G.S., of the Woodwardian Museum, when we found the new pits close to a farm some 500 or 600 yards east of the Spinney Abbey Farm.

The section was as follows:—

		ft.	in.
c	(3) Brown surface earth	1	6
	(2) "Head" of blue clay	0	9
	(1) Irregular gravelly zone, the pebbles being mostly flints and coprolites .... about	0	3
b	(4) Blue, yellow, and coarsely mottled plastic clay, with scattered coarse quartz and other sand grains, and numerous sandy concretions	2	0
	(3) The 'silt bed'—a chocolate-brown and yellowish sand passing into a sandy clay, which is rather coarse, loose, and like an ordinary shore sand. It consists principally of quartz and iron grains. A few irony sandstone nodules are scattered in the bed. This bed passes gradually into bed (4)	2	0
	(2) The "Upper Coprolite seam." A pebble bed of phosphatic nodules, Lydian stone, chert, quartz, and other pebbles as big as beans, packed in loose iron-coloured sand. Some irony concretions occur in its upper part, where it passes into bed (3)	2	0
	(1) The "Lower Coprolite seam." A thin band where the 'coprolites' are darker and better than in the upper seam. The sandy matrix is hardened almost to a 'rock' by carbonate of lime, which was probably derived from the underlying bed (a)	0	3
a	A calcareous grit of corallian age. It is a hard, gritty, bedded limestone; grey coloured, with scattered large oolitic grains; no fossils seen.		

The clay bed, No. 4, looks at first sight very much like the gault clay, but although beautifully ductile it is seen to contain an abundant scattering of large grains of quartz, ironstone, etc.; and many irregular nodules of coarse irony sandstone, as big as potatoes, are also found throughout its thickness. We have then no hesitation in referring this bed to the Lower Greensand—a conclusion which is further supported by its perfect transition into the 'silt' bed below.   ·

Also this clay bed yielded specimens of beautifully fibrous fossil wood, reminding one of the Lower Greensand wood of Shanklin, Isle of Wight.

The fossils of the Upware and Brickhill Neocomians are preserved, for the most part, in calcite, the mineral being, in some of the zoological groups distinctly crystalline; some few organic structures have been replaced by limonite, and fossil wood occurs in the usual silicified condition. Of other fossils only the inside moulds and external casts are known, these being formed of ferruginous sandstone, limonite, and phosphate of lime.

THE 'DERIVED' FOSSILS.

It is of the first importance at once to divide the fossils of these rocks into two great groups. All those which are mineralised in phosphate of lime, together with many of those in limonite are '*derived*' *fossils*. They are the remains of organisms which never belonged to the Neocomian period itself, any more than a fossil in flint now washed out from the Dover chalk and buried in the sandy shore belongs to our present epoch.

This separation of the fossils into two groups is, as a rule, as perfectly easy as it is a conspicuous necessity; for the *derived* fossils besides being characterised by the materials in which they are preserved, belong mostly to Jurassic species, and they occur usually as internal moulds which have been mutilated by attrition as they were rolled into pebbles during long years of wear and tear upon the ancient sea beach[1][2].

----

[1] Mr J. F. Walker first pointed out that at Potton there are two great groups of fossil remains, which are in this place (a) the indigenous fauna, preserved in Oxide of Iron, (b) the derived fossils, preserved in phosphate of lime.

[2] Mr H. G. Seely has given it as his opinion that all these fossils were natives of the bed in which they are found. *Annals and Mag. Nat. Hist.* 1866.

THE PHOSPHATIC NODULES.

The *phosphate of lime nodules* are mostly of small size, ranging from mere grains up to masses equal to the double fist. They consist of a dull compact material, the better kinds having a cubical fracture and bearing much resemblance to chocolate cake both in colour and texture. They are about as hard as ordinary limestone, or septarian nodules.

The colour varies from dark chocolate brown to a pale creamy yellow, these variations depending in part upon the nature of the original rock from which the nodule was derived, but not entirely so, since it is a general rule that the darker 'coprolites' are found towards the bottom of the pit. The darker nodules are said to be richer in phosphate of lime; other varieties of the phosphatic nodules are produced by the varying admixture of sandy matter in them, so that we get all the different degrees of purity from the perfectly homogeneous and compact phosphate, through specimens with more and more abundant sandy particles till we get a very impure phosphatic grit. Mr Walker has pointed out to me that many of the Portlandian species (*Cardium dissimile* and the *Trigoniæ*) are commonly found in a dark gritty variety of phosphate, but I am not able satisfactorily thus to separate out the phosphatised fossils into groups of different ages according to differences in the matrix.

The phosphatic nodules are either fragmentary portions of larger masses or the casts of fossils. In the latter case the exposed outlines are, as a rule, thoroughly worn into rounded surfaces of abrasion, so that the original shape is scarcely, or not at all, recognisable; they are also much tunnelled by numerous boring organisms. These stone-borers were for the most part sponges, worms, and bivalve shells, especially the latter, whose crypts often cover the whole surface of the nodule, leaving the outer zone of the 'coprolite' thoroughly honeycombed. But besides these there are other small markings of very general occurrence upon and throughout the phosphatic nodules whose nature is far more problematical ; namely, certain curious branching, interlacing, undulating or simply straight-crossing structures forming little gutters over the surface of the nodule, and canals penetrating into its substance. Some of these are mere shrinkage cracks and others are the marks of where 'episites' such as serpulæ and polyzoa have been attached to the inner surface of the original shell; others again are probably the work of boring creatures, especially sponges, but the great variety and the many peculiarities of type[1] that occur, and their constant association with phosphatic nodules are facts not sufficiently explained by the accumulated work of all the above-mentioned agents[2]. I have however no further explanation to offer.

An analysis of the coprolites by Voelkler may be consulted in Prof. Bonney's *Cambridgeshire Geology*, p. 25, where it will be seen that these "red coprolites" contain about 40 or 50 per cent. of phosphate of lime[3].

DISTRIBUTION OF THE NEOCOMIAN COPROLITES.

The phosphatic nodules are far from being limited to the areas around Cambridge and Bedford, where they have, however, alone been worked. Precisely similar ones are found in the Lower Greensand at Hunstanton, Ampthill, and Farringdon, in the pebble beds at Godalming and Redcliff (Isle of Wight), and in the Tealby section in Lincolnshire. In the last-mentioned locality they form zones of 'coprolites' along the junction lines of the Neocomian series both with the Kimmeridge clay below and the red chalk above[4].

The nodules of the Downham Market coprolite-bed are of quite a different character from these as described in a later page, and we shall find reason for believing this bed to be of a different age from the Upware rock. But crossing the N. Sea to the Brunswick area and the Harz of Northern Germany we again find at

---

[1] I have seen them in the Cambridge Greensand with all the beautiful symmetry of a fern frond.

[2] These structures are found in the Lower Greensand 'coprolites' of Upware, Brickhill, Potton, Farringdon and Schoeppenstedt (N. Germany), and in the more recent phosphatic nodules of Downham Market, the Cambridge Greensand, Buckinghamshire Gault, and the *Mammillaris* zone of the Ardennes chain.

[3] The 'Black Coprolites' of the Cambridge Greensand (above the gault) are much richer, containing about 50—60 per cent. of phosphate of lime.

[4] Since the time when I first found these nodules, five years ago, Mr A. J. Jukes Browne, of H. M. Geological Survey, has independently discovered the lower seam above mentioned, in several places, so as to demonstrate its value in stratigraphy.

Goslar, Schoeppenstedt, and Salzgitter, the same pale red, yellow and brown phosphatic nodules with the same phosphatic casts of *Ammonites* and *Myacites*, which could not be distinguished from those at Upware and Potton.

The contemporaneous Nodules of the Upware and Brickhill Nodule bed. These are utterly insignificant as compared with the ordinary derived nodules; they are not recognised as 'coprolites' by the workmen, and are rejected from the sorted material by the 'pickers'.

Looking over the waste rubbish heaps at the workings we occasionally meet with an irregular or cylindrical nodule of a hardened sandy nature, made up of quartz, ironstone, and phosphatic grains and fragments, and shewing, more or less distinctly, its phosphatic character on a fractured surface. The amount of phosphate of lime contained in them is very small and there may be none at all distinguishable by the naked eye; so that at Upware they never merit the title of Phosphatic nodules. They occur in the sandy beds, mostly near the junctions with the coprolite seams.

Contemporaneous nodules of the Neocomian period do however occur in very pure forms in other districts. At Speeton, besides the junction coprolite bed resting upon the Portlandian zone they occur scattered through the clay series, so that a number of the indigenous fossils in the shrimp-bed and elsewhere (*Meyeria, Ammonites,* &c.) are now permeated with and surrounded by the phosphate. Again in the Tealby series of Lincolnshire septarian phosphatic concretions are scattered through the clay-bed beneath the ironstone, being formed around large crustaceans (*Hoploparia?*) and other fossils as nuclei. In the Hanoverian clay of the Deister we again meet with the contemporaneous phosphate around *Meyeria*, or in precisely the same condition as at Speeton. But the most important occurrence of contemporaneous phosphatic nodules is at Downham Market, in Norfolk, first recorded by Mr Teall in his Sedgwick Prize Essay, pp. 20, 21. For here they are so rich and abundant that they are worked for manure.

In Dec. 1876, the section at Downham Market was as follows:

(4) *Gault* becoming sandy towards the base; containing scattered phosphatic nodules or "Militiamen," and phosphatised fossils. *Ammonites interruptus, Inoceramus concentricus,* etc., 10 ft.

(3) The coprolite bed; a yellowish sand sometimes hardened by iron oxide, in which the Phosphatic Nodules or 'Regulars' (as the workmen style them) lie. Fossils. About 1 ft. 2 in.

(2) A rock bed or 'carstone,' ironstained, 5 in.

(1) Red, white and green sands; proved by a well-boring to be 20 ft.

Both the "Regulars" of the true coprolite bed and the "Militiamen" of the gault are pressed into the service of manure manufacture, and are mixed up together in the heaps of nodules. The 'Militiamen' are of the usual gault type, but the 'Regulars' are different from any that we have yet noticed. These are rather large and rounded nodules, but unworn ; and we may distinguish two kinds amongst them (1) those of a dirty white colour, and (2) the dark green ones. Their surface is not smooth as with the much eroded derived nodules of other places, but rough and scabrous with projecting grains of quartz, &c., as though sand had been sprinkled upon a viscid hardening mass. This is especially the case with the pale forms.

Breaking a nodule to expose a fresh-fractured surface we see it is a quartzose, and even pebbly sandstone of varying coarseness, the grains cemented together by a dense matrix of phosphate of lime and iron oxide. The dark green nodules are of an older date of origin for they are frequently included within the paler ones. In internal structure the two types are similar, but the green ones are perhaps usually more compact, slightly darker in colour, and apparently richer in phosphate than the paler ones. The fossils of the Downham coprolite bed are principally in the state of internal casts in a dark variety of phosphate, more like those of the Cambridge greensand and the *mammillaris*-zone nodules of the south of England and the Ardennes[1].

Origin of the ' coprolites.'—It is a very general rule that phosphatic pebble beds are found along lines of ancient erosion and unconformity, the nodules having been derived from the destruction of older rocks. Thus the 'coprolites' of the Suffolk Crag are derived from the London Clay and those of the Cambridge Greensand to a great extent from the gault. Just similarly the nodules of Upware and Brickhill have been derived, for the most part, from

---

[1] The Gault nodules of the Perte du Rhône and the Bala Phosphates belong to the contemporaneous class.

the Upper Jurassic rocks when these beds were being denuded by the work of the Neocomian waters. The evidences of this are plainly written upon the nodules themselves, they being to a great extent the mutilated remains of Mollusca of Oxfordian, Kimmeridgian or Portlandian species. This being the case one naturally searches these formations for such a supply of Phosphatic nodules as might furnish an accumulation of 'coprolites' like that of the Upware deposit. Now phosphatic nodules *do* occur in the Kimmeridge clay, as indeed is the case with most of our Mesozoic clays, but they are everywhere very few in number, and the great majority of the Jurassic fossils are not phosphatised at all, this condition being most exceptional. We therefore at once meet with a difficulty in the way of such a theory. Moreover if the phosphatic nodules had been derived, just as they are, from the older rocks, we might reasonably expect that the nature and quantity of the phosphate of lime would vary in regular recurrence according to the deposit from which the species was derived, whereas we find that this is not the case. The general character of the phosphate of lime nodules is similar, whether it has come from the Oxford clay, Kimmeridge clay, Portlandian rock, or from the older Neocomian itself, and again *nearly all* the derived fossils are found in the condition of phosphatic casts. It is therefore my belief that although the phosphatised nodules and fossils were derived from older rocks, yet the phosphatic matter itself was all, or nearly all, of one age, namely contemporaneous with the deposit in which we now find them, that is Neocomian.

In the words of Mr Walker the "phosphatic nodules had been formed of clay [marl and limestone] soaked in decomposing animal and vegetable matter" and saturated with phosphates. And such is the theory adopted by Mr Teall in his Essay (p. 39)[1]. A similar case is that of the Ashley phosphates of South Carolina which, according to MM. Holmes[2] and Leidy[3], are sands and clays of

---

[1] The formation of the Cambridge Greensand coprolites, which are very different, has been very fully treated of by Prof. Bonney, in his *Cambridgeshire Geology*, p. 63, but I have long had suspicion that the amount of phosphate has been increased since they left their bed in the gault, because the nodules in the gault are generally softer and appear less rich in phosphate of lime.

[2] *The Vertebrate Rocks of S. Carolina*, by F. S. Holmes, A. M. Charlestown, 1870.

[3] "Vertebrate Remains, chiefly from the Phosphate Beds of South Carolina."— *Journ. Acad. Nat. Sci. Philadelphia.* Vol. VIII. part 3.

Post Pliocene age in which masses of more coherent rock, rich in calcium phosphate are intermingled. The masses are of irregular shape and of all sizes, "from pebbles up to 1000 pounds and more" (Leidy), and they are much tunnelled by Lithophagi. Now these masses, except for their phosphatic elements, are similar to the rock of the underlying Eocene and Miocene formations and contain the same fossils. Concerning their origin Prof. Leidy writes "at this time [when the older Eocene and Miocene Marls were being broken up] or later, neighbouring and superficial islets, the resort of myriads of sea fowl, may have furnished the material which when washed with the ocean and mingled together with the decomposing remains of marine animals, supplied the elements for the conversion of the porous rock marl into the more valuable phosphatic compound." Such was the explanation originally offered by Mr Holmes, and suggested, it seems to me, long ago by MM. Way[1], and Paine who also considered the phosphates to be of organic origin. As a result of their chemical researches these gentlemen write,—"That the phosphate of lime has penetrated the various fossils and nodules from without there scarcely exists the smallest doubt."

Some actual cases of such an addition of phosphatic material to calcareous rocks in very modern times have been described by Prof. Dana[2] as occurring in fragments of coral rock included in guano at Howland's Island in the Pacific ocean; and another interesting example observed in Cave Ha near Giggleswick, was recorded by Mr J. E. Marr of St John's College in the *Geological Magazine* for 1876, p. 268. From this account it appears that some stalagmitic layers of carbonate of lime have become phosphatised by the percolation through them of water which had passed through an overlying organic deposit, made up for the most part of the bones of small mammals and birds (from the pellets of owls, etc.). "The only apparent way," writes Mr Marr, "in which it seems that the travertine could have become phosphatised, is by the percolation of water through the pellet bed, until it was

---

[1] "On the Phosphoric Strata of the Chalk Formation," by J. M. Paine and J. T. Way.—*Journal Royal Agricultural Society*, 1 Ser. Vol. IX. pp. 56—84.

Mr Godwin Austen believed, 1848, that the phosphatic matter was of coprolite origin.—*Quar. Jour. Geol. Soc.* Vol. IV. p. 257.

[2] *Corals and Coral Islands*, p. 283.

stopped by the stalagmite, which would slowly absorb the water
and allow time for the chemical reactions which resulted in the
phosphatizing of a portion of it."

With such additional supporting evidence therefore I have no
hesitation in receiving the theory of Mr Holmes as the true explanation of the origin of the phosphate of lime in our Neocomian
nodules.

There yet remains however one remarkable fact left quite un-
accounted for by this theory, that is, the condition of the derived
fossils and fragments from the coral rag of Upware.

These limestone fragments are not in the least phosphatised,
but are quite in their original condition as now seen in the coral
rag quarries; and yet the fragments are imbedded well in the
heart of the coprolite bed. I can only suggest that this purer
carbonate of lime was uncongenial to the phosphatic matter, which
was taken up more readily by the more argillaceous fragments.

CHAPTER II.

THE INDIGENOUS FAUNA.

OF the proper denizens of the Upware and Brickhill Neocomian sea, we number 176 species, of which 151 occur at Upware, and 86 at Brickhill. Nearly all of these are well preserved in calcium carbonate, which in the Gasteropodous and Lamellibranchiate shells, and the Echinoderm tests and spines is distinctly crystalline. They are thus readily distinguished from the derived fossils.

The detailed account of these fossils occupies the greater part of the following pages, just as the work of their comparison, identification and description has formed the principal labour of this essay. We will however here give some general analysis and summary of the occurrences and development of the various types of life which characterise the deposit.

The most striking features of the bed are the magnificent developments of the Lamp shells, or Brachiopoda, and its exceeding richness in large cup sponges, and massive and dendroid Polyzoa.

The total number of species of *Invertebrata* is 161, and many of these prove to be new to science, as indeed we should expect to be the case in such isolated areas and such 'episodal' deposits as our Upware and Brickhill phosphatic beds.

NATIVE VERTEBRATES.

The remains of vertebrated animals are not nearly so abundant as in the Potton bed. Two ordinary cabinet-drawers would contain all the vertebrate remains from Upware and Brickhill in the Woodwardian Museum—a meagre series indeed as compared with the rich collections of reptile bones found in the Potton district, (see the record of Prof. Seeley, "Index to the *Reptilia*, etc., in the Woodwardian Museum," Cambridge, 1869).

An important difference of opinion exists as to the age of these vertebrate fossils, Prof. H. G. Seeley believing that they were all living in the Neocomian sea and the neighbouring lands, while Mr Walker regards them as being derived from the denudation of the Kimmeridge clay, Wealden, and other rocks. The latter is, at present, the generally received theory, but we shall, I think, find reason to believe that our Neocomian bones belong to both these sets, some of them being 'derived,' while a great many of them really lived during the period when their present rock matrix was being deposited.

Many of the fossil teeth, both of Reptiles and Fishes, are identical with well-known Kimmeridgian species, and yet they are often remarkably free from marks of wave-rolling.

My impression for a long time was that the vertebrate remains were, with few exceptions, derived fossils, but more recently the arguments of the general likelihood of such reptiles as the Iguanodonts, Crocodiles, and Saurians having lived on the shores and neighbouring land, together with the good state of preservation of the specimens, have led me to believe that many of them are proper to the bed in which we find them.—Still more weighty evidence is the occurrence of certain species (Sphærodus Neocomiensis, Pycnodus, etc.) in various other Neocomian areas, widely separated from one another, as, namely, Cambridgeshire, Bedfordshire, and Shanklin (I. W.) in Britain; the Perte du Rhône, Landeron and elsewhere in Switzerland; and Alais (Gard), and Auxerre (Aube) in France. This cannot be due to the mere accident of specimens being washed out of older formations, for the surrounding and underlying rocks in these places are quite different and unsuitable.

Again, these fossils are extremely rare in the Jurassic rocks (I have never seen a specimen of *Sphærodus gigas* from the Jurassic rocks of the Eastern counties), whereas in the Neocomian beds they are, some of them, quite abundant; nor is there any such correspondence in proportionate numbers and species in the two formations, the Neocomian and the Jurassic, as should surely be the case if the fossils of the one were derived from the other[1].

Lastly, this belief in the indigenous character of many of the vertebrates was greatly strengthened, and, indeed, became a conviction, when I again examined the lithological condition of the bones; for most of these are but slightly, if at all phosphatised, but are impregnated with iron oxide; and yet these are the very fossils which we should most certainly expect to be phosphatised, both because of the presence of phosphate of lime in their original composition and because of their many irregularities of surface which are just the kind of spots where phosphatic matter is most prone to accrete.

But in examining a large series of the bones, I find that some of them *are* phosphatised, and very thoroughly phosphatised too, with here and there little masses of phosphatic matter adhering. Examples of this may be seen in the Woodwardian Museum amongst the *Ichthyosaur* vertebræ; in the Pliosaurs (Case VI., a. 1—3), and in the *Plesiosaurs ;* (an example of the latter with adherent phosphate may be seen in v. b. 4). It is these, and these only, that are 'derived' fossils, washed out of the Jurassic clays.

I must here state that in an unassorted heap of the Potton bones the truly *derived* phosphatised bones would form a much greater proportion than in the cases of the Woodwardian Museum, for these being the most worn and least well-conditioned of the specimens, have been to a large extent picked out and rejected from the better collections.

Of the *Cephalopoda* we have but very scanty representatives from the Upware and Brickhill beds, but these species are of great interest and importance to us on account of their limited ranges in space and time. Thus the *Ammonites Cornuelianus* is characteristic of the Hythe beds of the south of England and of the Aptien of southern Europe, and *Am. Deshayesii* is equally an Upper Neo-

---

[1] Exx. At Ely, in the Kimmeridge Clay, *Gyrodus* and *Asteracanthus* are the most abundant fishes; other Ganoids, such as *Ditaxiodus*, &c., also occur, but *Sphærodus gigas* is unknown. On the other hand, in the Neocomian, *Ditaxiodus* and many other Kimmeridgian fish are unknown, and the *Sphærodus* teeth are some of the commonest fossils. The same argument is equally strong with the Reptiles.

comian species both at Speeton and Atherfield in England, and on the continent. But the *Belemnites* (*B. pistilliformis* and *B. sub-quadratus*) are commoner in lower zones (Middle and Lower Neocomian) at Tealby and Speeton. The new species *B. Upwaren-sis* is a very remarkable form.

Ancyloceras Hillsii, Sby. is characteristic of the Hythe beds in the south of England.

Specimens are rare of all the species of Cephalopoda, and they are somewhat more fragmentary than is the case with the other groups. The Belemnite guards preserve their original structures of fibrous carbonate of lime[1], and their upper ends present the powdery, decomposed appearance and the successive off-shellings of the subgenus *Actinocamax*. The Ammonite shells are changed into crystalline calcite, with one exception, which is a mould in yellowish sandstone. Only one Cephalopod is known from Brick-hill, namely *Belemnites subquadratus*.

Of *Gasteropoda* there are 16 species, six of them being small turbinated shells, prettily ornamented with ribs and granules. Most of the univalves are undescribed species, and consequently of little value in stratigraphical comparisons. The *Pleurotomaria gigantea* is however a well-known shell in the Hythe beds of the Wealden area, and in the Isle of Wight Greensand; and the large *Nerinœa* shells are of interest as recalling the French Neocomian types. All the shells are now made up of crystalline calcite with the exception of the large *Pleurotomaria*, which is an inside cast in coarse sandstone.

At Brickhill no gasteropodous shell is known, but the markings of a species of Trochus (?) are found impressed in the base of a polyzoon, *Multicreseis Michelini*.

Amongst the 54 species of *Lamellibranchiata* the *Ostreidœ* are conspicuous, as presenting the most striking forms and being the most numerous in individuals. Such are *Ostrea frons*, var. *macroptera*, *Pecten orbicularis*, var. *magnus*, *Neithea ornithopus*, Kpng. and the little *Plicatula Carteroniana*.

The great resemblance of the Upware oysters to those of the Jurassic rocks is very remarkable. A large expanded Gryphoid

---

[1] At Potton specimens occur changed into peroxide of iron.

species is undistinguishable in form from the Oxfordian *Gryphea dilatata;* another shell (a form of *O. Walkeri*) has long been labelled *Ostrea deltoidea* in the Woodwardian Museum, and it is indeed precisely like it; and, again, the little *Exogyra Tombeckiana* (called *E. nana* in the earlier lists of Mr Walker) seems to pass into the *Exogyra nana* of the Kimmeridge clay. The Brickhill oysters are somewhat different, *O. macroptera* being there represented by *O. carinata.* It is not inopportune to point out the close resemblance of this latter species also to the *Ostrea gregaria* of the upper Jurassic rocks.

The special development of the *Arcadæ* at Upware is also noteworthy, and particularly the occurrence of several species of the uncommon genus *Pectunculus.*

The *Opis Neocomiensis* and *Cyprina Sedgwickii* are conspicuous and beautiful fossils at Upware, and the isolated appearance here of several species of such a peculiar group of *Cypricardia* (*C. striata, C. arcadiformis,* and *C. squamosa* Kpng.) is very remarkable; the more so because one of them is identical with a species described by Prof. Geinitz from the far-off shore line of the upper chalk ocean at Dresden.

Small boring bivalves of the *Modiola* group are abundant, the shells being still in their crypts.

At Brickhill the Lamellibranchs are much less common than at Upware; only fourteen species are known there, and eight of these belong to the *Ostræidæ.*

BRACHIOPODA.

The Upware Brachiopoda have been so thoroughly worked out in the published works of Mr Walker and Mr Davidson that it is not necessary here to describe them with the same detail as the other groups. But since the publications of these authors appeared the new workings at Brickhill have yielded us much fresh material for work, including many new and interesting forms and some additional species.

Both the Upware and Brickhill Neocomian sea bottoms were particularly rich in the 'lamp shells.' They grew so thickly

clustered together in places that, being overcrowded, they inter-
fered with each other's proper growth and produced abnormal
distortions[1]. Brickhill was the metropolis of the Brachiopoda, in
Cretaceous times; but still, although the species are so numerous
(35 in all), the genera are few, namely, only *Terebratula, Waldheimia, Terebratella, Kingena* (1 sp.), *Terebratulina* (1 sp.), and
Rhynchonella. Amongst these the *Waldheimia Woodwardi* and *W.
pseudojurensis*, and the species of *Terebratella* are the most noteworthy as remarkable and characteristic types. Of the *Terebratellæ* only two species *T. Meyeri* n. sp. and *T. Davidsoni* are common
at Upware; no well-marked representative of the other species
having to my knowledge occurred in that neighbourhood. *Terebratula capillata* is remarkable on account of its limited and curious
distribution, being only known in these Neocomians, the Red chalk,
and the Tourtia of Belgium. *Terebratula Upwarensis, T. microtrema,
T. prælonga, T. depressa* and *T. Montoniana* were extremely abundant,
but *T. Meyeri, Lankesteri* and *Dallasii* are rare. At Brickhill the
Brachiopods occurred in the richest profusion, and a noble series of
them may be studied in the Woodwardian Museum. The near
relationship of the Brickhill species to the Upware group is incon-
testable, all the Upware species except *Waldheimia Woodwardi,
Terabratella Meyeri* and *T. capillata* having been found here also.
The most noteworthy points of difference are the much greater
development of the Terebratulæ at Brickhill (*T. Keepingi* is unknown
elsewhere), the presence of *Kingena rhomboidea* and *Terebratulina
striata*, and the absence of the species aforementioned. It may be
observed that these Brickhill species, absent from Upware, are
more southern forms, known at Farringdon, or in the Wealden area
at Godalming and Hythe. The tables of distribution are, in the case
of the Brachiopoda, built up almost entirely from the authority of
Mr Davidson, and Mr Walker, Mr Meyer and M. Barrois, or from
my own comparisons; for the continental records are in such confusion as to be quite untrustworthy.

[1] Mr Walker possesses a specimen of *T. Meyeri* in which the irritation produced by the working of a boring shell has caused a pearl to be formed in its
interior.

*On Varietal and Intermediate Forms amongst the species of
Brachiopoda.*

It is a matter of general experience in working amongst the
more prolific families and genera of organic types that the boundaries of our recognised species are found to be most shadowy and
evanescent. In any particular species we have the well-marked
typical specimens, but we also meet with a number of 'intermediate
forms,' specimens 'not well marked,' with the specific characters
ill-defined, and approaching an allied type. Unfortunately these
specimens are, as a rule, rejected by collectors as troublesome and
unworkable, and are therefore stowed away in boxes amongst the
duplicates and lost sight of. This is a most regretable custom, by
which we have already lost a mass of valuable material. The
value of the work of separation of species and the discrimination of
their special characteristics are not likely to be underrated by the
Geologist and Naturalist, but the discovery of their relations to
one another and the determination of the positions they have
occupied in the development of the great life groups is likewise
a noble work, and a work not only of profound interest to the
Biologist, but one that we may also hope to be of great value even
to the Geologist in determining the relative ages of rock-beds.

The acquisition of large numbers of Brachiopods from the rich
collecting grounds of Upware and Brickhill have furnished us with
unusually good opportunities for working out the meaning of these
variations amongst the species of *Terebratula*, &c. At Upware
many of the species are well defined, as *Terebratula Upwarensis,
Waldheimia Woodwardi, Terebratella Meyeri*, and *T. Davidsoni;*
but some species are so freely connected that an ordinary small
collection cannot, to our satisfaction, be completely separated out
into its several species. Such are, for example, *Terebratula depressa,
T. prælonga*, and *T. microtrema*. But it is at Brickhill that these
intermediate forms are most conspicuous and important; from this
locality the Woodwardian Museum has accumulated upwards of
15,000 specimens of Brachiopods, and from these I have selected a
series of specimens and arranged them upon tablets, to shew the
relations of the species to one another, as shewn by intermediate
connecting forms, in the following manner:—

Genus Terebratula.

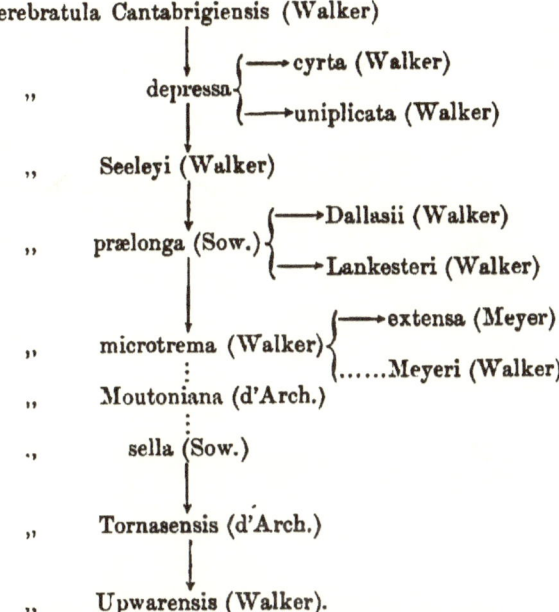

Terebratula Cantabrigiensis (Walker)

     „   depressa   ⟨——▸cyrta (Walker)
                         ⟨——▸uniplicata (Walker)

     „   Seeleyi (Walker)

     „   prælonga (Sow.)   ⟨——▸Dallasii (Walker)
                              ⟨——▸Lankesteri (Walker)

     „   microtrema (Walker)   ⟨——▸extensa (Meyer)
                                ⟨......Meyeri (Walker)

     „   Moutoniana (d'Arch.)

     „   sella (Sow.)

     „   Tornasensis (d'Arch.)

     „   Upwarensis (Walker).

Explanation. The arrows indicate the lines of variations and probable development. The dotted lines signify less perfect continuity and connexion.

This series may be studied in the drawers of the University collection, Cambridge.

Amongst all these species the connecting morphological varieties are by no means equally evident or of equally frequent occurrence. Between *prælonga* and *microtrema* the connecting varieties are very abundant, so that in any dozen specimens, casually taken, one is pretty certain to meet with the difficulties of separation; and indeed through all the linear series in the plan from *T. Cantabrigiensis* to *T. microtrema* the passage is simple and clear. Again, the varietal offshoots from *depressa* (*cyrta* and *uniplicata*), and the lateral branches from *prælonga*, known as *Lankesteri* and *Dallasii*, are satisfactorily connected with the main stock. But much more searching and much greater care is required to establish the series between *T. Meyeri* and *T. Moutoniana*; nor is the result so satisfactory as in the other cases; it is in fact remarkable that it proved to be much more difficult to arrange a series between these two somewhat similar shaped shells than between the far more diverse *microtrema* and *depressa*. In the original description of *T. prælonga* and *microtrema* it will be found that their different microscopic shell-characters are much insisted upon as serving to separate these two species from one another; the shell tubules in the latter species being much smaller than usual, whence the specific name was given: but as with the other characters so here, there is every intermediate variety, and indeed I find both types of structure in the same shell. Thus every character fails in an attempt to find definite and unvarying points of distinction between the several species.

The appended Table is the result of the same kind of work amongst the long-looped genus:

WALDHEIMIA.

Waldheimia celtica (Morris)

| |

„   Wanklyni, var. angusta (Walker)

| |

„   Wanklyni, var. elliptica (Walker)

| |

„   faba (d'Orb.) (?)

| |

„   pseudojurensis (Leym.)———►many varieties———►new species
(*Hughesii* Kpng.)

„   tamarindus (Sow.)

| |

„   Juddii (Walker).

In this table all is plain sailing from *W. celtica* through the forms of *Wanklyni* to the flattened varieties of *pseudojurensis*, and no one will dispute the continuity of the series through the many varieties of the last-named species, with *digonal* and elongate forms, inflated and rounded, or thin and flattened, and including the new type, *W. Hughesii* (Keeping). From this point the chain of passage to *Waldheimia tamarindus* is somewhat strained, and is perhaps not really true in nature, but from the latter species we pass on again without break or difficulty to arrive at *Waldheimia Juddii* (Walker).

The *Waldheimia Woodwardi* (Walker) remains isolated, belonging as it does to quite a different type from any other of the Upware and Brickhill species. It is the extreme form of the *hippopus* group, whose nearest occurrence is at Tealby, in Lincolnshire. The plan

Waldheimia hippopus——▸ *W. Walkeri*——▸ *W. Woodwardi*

may indicate the line of evolution of this species.

The species of the *oblonga* group of *Terebratella* fall into line with perfect ease in the order indicated in the following table:

TEREBRATELLA. GROUP A.

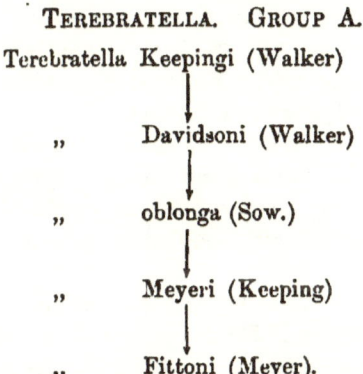

Terebratella Keepingi (Walker)

„ Davidsoni (Walker)

„ oblonga (Sow.)

„ Meyeri (Keeping)

„ Fittoni (Meyer).

and moreover I find such variations amongst the allied continental species that it is clear to me that with the help of a good measure of specimens these also could be readily adjusted to fit into our developmental series. Thus, we have in Germany, *T. oblonga, Puscheana* and *Beaumonti*, and in France and Switzerland, *T. Meyeri* [?] and *semistriata*. This group of *Terebratella* is markedly characteristic of the Lower Cretaceous rocks all over Europe.

Of the straight hinged types of *Terebratella* transition forms occur which bind well together the two species

T. Menardi

T. trifida,

but we have at Upware and Brickhill no traces of any passage of the two types *oblonga* and *Menardi* into one another.

After having thus studied our materials with respect to their interrelationships, it becomes a matter of special interest to inquire into the general results and their meanings, and the interpretation of those points of difficulty where our chains were weakest.

Now our most troublesome points were in the connexion between *Moutoniana, microtrema,* and *sella ;* between *microtrema*
and *Meyeri ;* and between *Waldheimia pseudojurensis* and *W.
tamarindus.* Now if from the groups of *Terebratula* and *Waldheimia* we select all the species that are known in rocks of older
date than our Midland Neocomians, we shall have *Ter. Moutoniana,
T. depressa* and *T. sella, Waldheimia tamarindus* and *W. pseudojurensis,* all of them occurring in the Middle Neocomian of other
districts, and some in the Lower Neocomian. Now these are precisely the species where we have failed to establish a perfect series
of connecting links,—where our chain seemed weak or broken[1].
These species had acquired their special characteristics of form in
an earlier period and remained distinct.

The same was doubtless true of the Terebratellæ, for these fall,
not into one group but into two, and these two types must have become differentiated and fixed before the period of our Brickhill
deposit, and so of course remained distinct as separate local main-
stocks.

It is noteworthy that at Upware, and indeed all other places
known to me, the species of Brachiopoda maintain much more dis-
tinctness and isolation from one another than at Brickhill. And
from this fact one might expect to find that the Brickhill fauna
flourished at a slightly earlier date than that at Upware, and
acquired diverse developments into well-marked types ; and these
spreading outwards from this centre of dispersion to other areas
(Upware, Potton, etc.), afterwards maintained their special developments of form with all the fixity which commonly obtains in
specific characters.

In examining the gradational series arranged on tablets in the
Woodwardian Museum it should be borne in mind that these
tablets do not pretend to be complete in their every connexion at
each point. Indeed it is scarcely possible to arrange a perfect
lineal series. The characters to be considered are very numerous
in each species, and, in the absence of the ideally perfect connecting specimen, intermediate in all its points, it must suffice in

---

[1] It is worth recording that the arrangements here described had been completed and the points of difficulty here noticed were recognised before the idea of
their connexion with the earlier appearance of certain species had suggested itself
to me.

places to consider here one character and there another, and thus demonstrate the variability of each character by itself. Thus the series, though made somewhat more cumbrous, yet serves scarcely the less to illustrate a conviction which steadily and powerfully grows upon one in the course of many years' constant familiarity with a multitude of specimens.

While thus pointing out the mutability of the forms of Brachiopoda it is perhaps worth while observing that the value of the species is thereby in no way decreased, but on the other hand is, I believe, considerably increased both to the Naturalist and Stratigraphist.

Whether we call our various forms *varieties, races, types* or any other name, the facts of the great constancy of our recognised 'specific' types, and their limited distribution in space and time still remain to us: our characteristic species are as useful as ever they were; while on the other hand we shall have added a most important help to the determination of the relations of rock beds to one another when we can recognise the true meaning of allied genera, species and varieties.

Nor can I leave this part of the subject without referring to those false specialists of late years who, to shield their own half-knowledge of a subject, seem to be content in making loud professions of their belief that "there are no such things as species." These persons, it seems to me, are ignoring the house because the scaffolding has been taken away. They fail to see that the old values remain unbroken, when viewed in the new lights, and still less are they able to perceive the added virtues. Nor can they appreciate the laborious training of the modern Naturalist and the increasing importance of his work.

POLYZOA. The Polyzoa of Upware and Brickhill are remarkable for the rich development of the large arborescent forms. The *Ceriopora (Echinocava) Rawlini* is a peculiar coralloid type, originally described from the Ardennes but not uncommon at Upware. *Reptomultisparsa Haimiana* is a Swiss species, but the *Heteropora coalescens* is a North European form described by Reuss from the Unterer Quader (= Upper Chalk) of Saxony, near Dresden. The great amorphous, nodulose masses of *Ceriopora* are amongst the most striking of the Upware fossils; but of the ordinary types of creeping and encrusting Polyzoa scarcely a trace has been found.

The full number of species known is 26, 18 from Upware and 8 from Brickhill.

ANNELIDA. Amongst the Worms fine species of the Genus *Serpula* are common, and the *Vermicularia Phillipsii* occurs both at Upware and Brickhill. The *Serpula articulata* (Sowerby) is the representative of the *S. vertebralis* of the Oxford Clay.

ECHINODERMATA. The Echinoderms are few but interesting. A single specimen of *Peltastes Wrightii* has occurred at Upware, and at Brickhill it flourished in good numbers, but its full development was realized still further south, at Farringdon. The *Pseudodiadema* are fine and abundant from Brickhill.

SPONGES. This group was nobly represented in the Neocomian formation at Upware. Beautiful cup sponges of the group *Catagmidæ* (the old *Manon*) flourished around the Upware Coral Bank, belonging to the genera *Catagma, Elasmostoma, Verticillites* and others ; altogether they number eight species. Eleven species have occurred at Brickhill. The Upware sponges were remarkably limited in their distribution, all the good specimens having been found within the area of some two or three fields ;— unless indeed it be that they have been destroyed elsewhere by the percolation of water, which is improbable.

The curious septate and siphuncled sponge *Verticillites* reached its greatest development in these deposits, three distinct species (two of them new) having lived at Upware ; and the strange-looking fragments of the coarse mesh-work of *Pachytiloda* are also worthy of note. At Brickhill most of the species are the same as at Upware, but two new forms appear with narrow elongated cloacal cavities, which I refer to the genus *Peronella* (Zittel).

All these sponges belong to the group of *Vermiculata* (O. Schmidt), the "Schwamme mit Wurmförmigen Gewebe" of Roemer, so named from the vermiculate appearance of the anastomosing fibres of which these sponges are composed ; but at Brickhill another species belonging to the beautiful group of *Hexactinellidæ* or *Vitrea* has also been found, namely *Plocoscyphea pertusa*, a species described by Geinitz, and occurring in the Quader Sandstein, near Dresden.

The classification of sponges is now undergoing a complete

reformation in the hands of Professor Zittel, in Germany, and MM. Carter and Sollas, in England.

The old system of founding genera and families upon such characters as the simple or compound habit of the sponge, the openness or closeness of the cup, presence of epitheca, or according as the oscular openings were on the concave or convex surface, has resulted in the present utter confusion of nomenclature. Such characters were, indeed, often not even of specific value.

Unfortunately the modern reformers are not yet in harmony with respect to the nature of the group of sponges to which most of those at Upware and Brickhill belong. All the sponges from these beds (as also from Farringdon) now consist of carbonate of lime. Every fibre as now found is made up of that material, and Professor Zittel, having regard to this fact, and attaching especial importance to the occurrence of triradiate spicules in the fibres, regards them as having been likewise originally calcareous sponges, as they are now, in spite of their being so different from all the living representatives of the *Calcispongidæ*. Mr Sollas, on the contrary, giving more weight to the preponderance of simple bacillar spicules in the fibres, argues that all the spicules were originally silicious. That such a replacement of silica by carbonate of lime has often occurred in Nature, especially amongst the sponges, is recognised on both sides, so that the dispute affects only this particular group. In this connexion the discovery of the silicious hexactinellid sponge *Plocoscyphea* is of especial interest, for no one will contest the original flinty composition of this sponge, whilst now its most delicate structure is beautifully preserved entirely in carbonate of lime. Thus Mr Sollas' theory of substitution receives some special support from our Brickhill fossils; but we must leave this question to be settled as time goes on and new experience is gained.

We can however no longer rest content with the old methods of classification, but must in future base our classification firstly upon the nature and form of the constituent fibres and spicules, using only with great caution the features of external form, intramural canals, and surface oscula as generic characters. "The microstructure of the skeleton alone is of decided importance for the determination of all sponges," are the words of Professor Zittel[1].

---

[1] Translated by Mr Dallas, *Ann. and Mag. Nat. Hist.* 1879, p. 366.

CHAPTER III.

THE 'DERIVED' FOSSILS.

THE remains of fossil organisms which have been derived from the destruction of older rocks are very numerous in all the coprolite pebble-beds of Cambridgeshire and Bedfordshire. Referring to my notes on some pits near Ampthill I find "the coprolite heap looks like one mass of *Ammonites biplex*, mostly worn and fragmentary," and in all cases a large proportion of the coprolites shew traces of the outlines of shells.

The vast majority of the derived fossils are preserved in phosphate of lime, but the Ammonites of the Oxford clay are composed of limonite and some of the fragments of fossil-wood are silicified. The vertebrate fossil remains are more or less impregnated with oxide of iron or phosphate of lime, but we find some difficulty in deciding exactly to what extent these are derived, and which are native to the deposit.

The Coral Rag fossils from the neighbouring rock at Upware have not been phosphatised at all, but retain their original character although they occur well in the nodule-bed amongst a mass of other phosphatised fossils.

The proper shells of the derived fossils are usually lost, and the casts themselves have suffered much from trituration on the old sea shore, the burrowing work of boring shells, &c.

The present condition of these phosphatised fossils will best be understood from a description of a typical example. Two of the commonest are *Ammonites biplex* and a *Myacites*. The former occurs as a discoidal mass of oval or elliptical shape, like a mineral concretion, but in it one or more whorls of the Ammonite may here and there be marked in shadowy outline. No fragment of the shell itself is left upon the exterior, and only faint traces of the ribs of the inside cast remain. The umbilicus is filled with an irregular mass and protuberant lumps of phosphate of lime, often exhibiting fragments and impressions of other fossils. No abrupt shell-aperture now breaks the contour of the body, all this having been worn down flush with the general outline. Thus the action of the waves has changed the beautiful *Ammonites biplex* into an unsightly mass; and the work of mutilation has been further carried on by crowds of worms, sponges(?), Lithodomi, *Arca*-like shells, *Pholadidea* and such rock-boring animals, whose empty crypts now stud the surface of the relict of the Ammonite and its adherent phosphate. But within this mass the inner shell-whorls have been protected and saved from destruction by the older enfolding coils; and often, under a stroke of the hammer the central part will jump out in all its original freshness, a brilliant new birth glistening, it may be, with a charming play of iridescent colours from its pearly shell layer.

The other example is a bivalve shell of the *Anatinidæ*, commonly referred to *Myacites*. Here the shell is totally lost, and the original outline of the interior phosphatic mould has been to a great extent destroyed by wave-action, the umbo, pallial border, and all the margin being completely denuded into rounded outlines.

The Upware phosphates are smaller and very much more worn than those of Bedfordshire.

The derived fossils belong to various ages, ranging at least from the Neocomian to the Oxford clay inclusive.

I. THE 'DERIVED' NEOCOMIAN FOSSILS.

A. *The Phosphatised Forms.*

One of the most surprising facts connected with our Upware deposit is the occurrence of undoubted Neocomian species as derived fossils. *Ammonites Deshayesii* and two species of *Ancyloceras* occur at Upware in as thoroughly phosphatised and battered conditions as the *Ammonites biplex* and other species of the inferior Jurassic rocks. In both cases the shells have been fossilized in their original sediments and washed out of the old sea cliff or sea bed, just as fossils are now washed out and scattered along the beaches of our present sea lines. There they were broken into fragments and rolled into pebbles till they found a resting-place and were buried in the accumulating sediment together with the true aboriginal shells that had lived in the Neocomian sea. Lithologically I am not able to distinguish the derived *Ammonites Deshayesii* of the Neocomian from the *Am. biplex* of the upper Jurassic.

The following is a list of the phosphatised Neocomian derived species: *Ammonites Deshayesii*, Leym.; *Ammonites*, smooth species; *Ancyloceras*, sp.; *Hamites*, sp.; *Gervillia linguloides*, Forbes? *Thetis minor*, Sby.; *Terebratula ovoides*, Sby.; *Terebratula ovoides*, var. *rex*, Lankester.

At Potton we meet with the same list of species with, in addition, fragmentary phosphatised specimens of *Ancyloceras gigas*. In the Töck of Heligoland similar, though usually smaller, fragments of the same *Ancyloceras* are found upon the shore. But the richest of these derived phosphatised Neocomian faunas is that of Hunstanton, where the following species were collected for the Woodwardian Museum: *Perna Mulleti* (some of them derived, but some look like natives); *Pleurotomaria; Ammonites Cornuelianus* Forbes; *Ammonites Martini* (Forbes); *Ammonites Deshayesii* (Leym.); *Ammonites*, sp.? (allied to *Koenigi*); *Ammonites*, sp. 2; *Ancyloceras gigas* (Sow.); *Ancyloceras* (tuberculated species); *Nautilus*, sp.

B. *The Dark-grit Types of Derived Neocomian Fossils.*

A peculiar, dark-coloured grit rock, containing a special fauna, occurs at Upware in the form of derived pebbles and boulders. The same rock is found in precisely similar condition in the Potton sands of Bedfordshire and the lower red sands of Hunstanton, in Norfolk.

The Upware and Potton blocks may be described as hard, ragged-looking lumps of most irregular form but with eroded surfaces: the unevenness is produced by the fossils, which here project into angles, and there leave hollows where they have fallen away. It is a moderately fine-grained stone, the constituent particles being quartz and ironstone, cemented together with iron oxide. It may be to some slight extent impregnated with phosphate. Generally it is crowded with casts of fossils. The Upware blocks were small, and yielded the following species: *Cerithium granulatum,* Phill.?; *Solarium Neocomiense,* d'Orb.?; *Trochus* sp. (also at Shanklin, M.); *Trigonia Vectiana,* Lycett; *Trigonia,* sp.; *Mytilus lanceolatus,* Sby.; *Thetis minor,* Sby.; *Perna Mulletti,* Sby.; *Perna Ricordeana,* d'Orb.; *Pecten orbicularis,* Sby.; *Cucullæa errans,* Keeping; *Cucullæa Donningtonensis,* Keeping. There is good reason to believe that *Terebratula ovoides,* Sby. and *T. ovoides* var. *rex,* Lankester, likewise belong to this fauna, although at Upware they are commonly found in a thoroughly phosphatised condition.

So crowded are the fossils in this rock that one small block of it from Upware, now in the Woodwardian Museum, exhibits specimens of *Pecten orbicularis,* Sby., *Thetis minor,* Sby., a small triangular bivalve, and *Trigonia Vectiana,* Lyc. At Potton these blocks have been found with many of the same species, and in addition, *Fissurella Neocomiensis,* d'Orb.; *Pecten striatocostatus,* Goldf.?; *Lucina; Leda; Cardium?* small species; and Fossil-wood, with the woody fibres beautifully seen.

At Hunstanton large masses of exactly similar rock as big as large cannon-balls are found in a zone beneath the carstone and above the clay (see Wiltshire, *Quart. Journ. Geol. Soc.* VoL XXV. p. 188).

From these blocks we have in the Woodwardian Museum the following species: *Hamites* or *Ancyloceras,* small species with a double row of spines along the back; *Scalaria; Tornatella; Solarium Neocomiense,* d'Orb.; *Pecten orbicularis,* Sby.; *Nucula; Lucina; Cardium subhillanum,* Leym.; *Cytheræa Orbigniana; Goniomya Rauliniana,* d'Orb.; *Scrobicularia phaseolina,* Phill.; *Corbula;* and *Pholadomya,* and other species from the same rock are probably included in the list given by Mr Etheridge in the paper by the Rev. T. Wiltshire already cited.

The similar rock above referred to as occurring in the Isle of Wight is found in the upper part of the Lower Greensand at Horseledge, Shanklin, and at Blackgang Chine, and here it contains, like the Upware blocks, the casts of *Thetis minor.* The palæontological resemblances of these two rocks are however not so great as the lithological, for out of thirty species only seven are known to be common to the two. The fragments of fossil-wood are much alike in all these places.

Again, a very similar dark grit occurs in the Uppermost Lower Greensand at Folkestone, containing *Thetis minor* and other small bivalves, *Mytilus* and fossil-wood.

K.   3

The subjoined table gives a list of all the fossils known from the Neocomian Black Grit rock, with some notes on their distribution elsewhere.

FOSSILS FROM THE DARK GRIT BLOCKS.

Ammonites sp. (collection J. F. Walker) ... ... U.
Hamites, or *Ancyloceras,* small species with double row of
  dorsal spines ... ... H.
Emarginula Neocomiensis, d'Orb., of the Shanklin and Ather-
  field beds ... ... P.
* *Trochus,* small ornamented species ... ... P.H.
Trochus, also at Shanklin, Meÿer ... ... U.
Scalaria ... ... H.
Tornatella ... ... H.
Cerithium, muricated species ... ... U.
Pecten orbicularis, Sby., Hilgay (Teall) ... ... H.
  „   *striato-punctatus,* Roemer (Wiltshire) ... ... P.H.
* *Perna Mulletti,* Desh. ... ... U.H.
* „ *Ricordeana,* d'Orb. ... ... U.
Nucula ... ... H.
Leda ... ... P.
* *Cardium subhillanum,* large form ... ... U.H.P.
  Also occurs in the Donnington sands (Tealby), Shanklin,
  and Punfield (Meÿer).
Small triangular bivalve ... ... U.H.P.
* *Thetis minor,* Sby., of Shanklin ... ... U.P.
* *Cytherœa Orbigniana,* Forbes, of Shanklin ; and the Crackers,
  Isle of Wight ... ... H.S.
Lucina ... ... U.P.H.
* *Goniomya (?)*
Pholadomya Raulimana, d'Orb. of the Albien, Ardennes; also
  at Nutfield (Meyer) and Tealby ... ... H.
? *Scrobicularia phaseolina,* Phill. ... ... H.
Pholadomya ... ... H.
Panopœa Neocomiensis, Leym.
Corbula ... ... H.
* *Cucullœa errans,* Kpng., also at Herrimere, and (?) in the
  Tealby series at Claxby ... ... U.
* „ *Donningtonensis,* Kpng., also in Lower Neocomian,
  Tealby district ... ... U.
* *Mytilus lanceolatus,* Sby. ... ... U.
Trigonia spinosa, Park. ... ... U.
* „ *Vectiana,* Lycett, of Shanklin and Atherfield ... U.
* [*Terebratula ovoides,* Sby. var. occurs in the Herrimere rock,
  and is found phosphatised at Upware, Brickhill and
  Potton] ... ... U.P.

  U indicates the dark grit blocks at Upware.
  P   „   „   Potton.
  H   „   „   Hunstanton.

On certain Boulders of Neocomian Grits which occur in the East of England.

Some clue to the origin of the dark fossiliferous grits which we have just been considering, may, I believe, be obtained by studying the boulders of green grit which have been found in various localities on the surface in the gravels of the east of England. So long ago as 1812 this rock was made known in the description of *Terebratula ovoides* by Mr J. Sowerby from "blocks of sandstone containing greensand" (*Mineral Conchology*, Vol. I., p. 227, pl. 100), found in "some parts of Suffolk." I have found the same rock-boulders containing abundantly the same fossils at South Willingham, in Lincolnshire, and near the inn at West Dereham, Downham Market, Norfolk. The last-named block yielded, besides the *Terebratula ovoides* (Sby.) a *Lucina* and *Pecten orbicularis* (Sby.). Mr Teall in his Essay (pp. 22, 23) cites other localities, namely, at Southery, and in a pit at Hilgay, Norfolk, where he found several such blocks of sandstone with the fossils "*Panopæa Neocomiensis*, *Pecten orbicularis*, and *Terebratula depressa.*" And Mr E. Ray Lankester also mentions a gravel-pit at Thorpe, Suffolk, another at Snape, and the Drift at Stow Bardolph.

I have lately found that several large boulders around Cambridge belong to the same set. The great corner-stone by the side of a cottage close by the Coton public-house contains good specimens of *Pecten orbicularis* and *Belemnites subquadratus*, some of the latter very large; and from other boulders in the farm next to the rectory at Hardwick I obtained *Pecten orbicularis* and *Pinna*.

But the most important of these boulders (as I believe it to be), was dredged up from the bed of the Cam some few years ago at a place called Herrimere between Upware and Ely, where it was found by Professor Seeley and Mr Earwaker, and described by Mr E. Ray Lankester in his paper on "A new large Terebratula occurring in E. Anglia," in the *Geological Magazine* for May, 1870, p. 410. In this paper we are told that the matrix is "a fine sandstone conglomerate, closely resembling the matrix from Thorpe and Snape, and having small black pebbles scattered through it." The specimens are in the Woodwardian Museum. It is a curious grey sandstone, full of fossils, and the contained pebbles are Lydian stones and rolled phosphatic fragments which have been much penetrated by lithophagous mollusca. Altogether the pebbles are very similar to those of the Upware Neocomians.

Terebratula ovoides, Sby., and other fossils, were found by Mr Seeley.

Thus in all these localities extending from S. Willingham in Lincolnshire to Cambridge and Suffolk we have a set of similar rock boulders containing *Terebratula ovoides* and *T. rex.* But the Herrimere mass is of particular interest, because besides the *Tere-bratula* it contains a number of other species which serve to fix the geological age of the original deposit. These are, *Belemnites,* sp.; *Terebratula ovoides* (Sby.); *Cucullæa errans* (Keeping); *Pecten striato-costatus* (Roemer); *Pecten orbicularis* (Sby.); *Lima longa* (Roem.)?; *Pholadomya; Lucina* (same species as at W. Dereham); *Trigonia* (Mr Ray Lankester); *Exogyra Tombeckiana* (d'Orb.); *Avicula macroptera?* (umbo large, ribs stout and rounded).

In the notice of this rock in the *Geological Magazine* above referred to the importance of its fossils, as helping to settle the age of the grit boulders, was fully recognised, but I cannot find any proof that the rock occurs "*in situ* beneath the bed of the Cam," as there stated on the authority of Mr Seeley; nor can I admit that the fossils "agree with the stratigraphical position of the bed as the very highest of the oolites." Notwithstanding the long-continued work of dredging in the bed of the Cam, the rock has, I believe, never presented itself but on this one occasion, and as to its stratigraphical position a glance at the map will show that it lies between the coral rag of Upware and the Kimmeridge clay of Ely. Again, its fossils, such of them as give any evidence at all, point to the Neocomian age. Six of the species are, I believe, cretaceous, namely, those marked with an asterisk in the list.

Now of the Herrimere boulder fossils the two most special and characteristic species also occur as derived fossils in the Upware coprolite bed, namely *Terebratula ovoides,* and the *Cucullæa errans* of the dark grit; and the latter species is unknown in any other rock. I believe therefore that both sets of boulders, those now imbedded in the lower greensand of Upware and those scattered about the country in gravels and surface drift, were all derived from the same parent rock.

Other links in the chain of evidence are afforded by the second species of *Cucullæa, C. Donningtonensis* n. sp. and the large *Cardium subhillanum* from the Upware dark grit fragments, for both of these are found in the Lower Neocomian sands of Donnington, in Lincolnshire. At this place, scattered through the lower green-coloured sands (Lower Neocomian of Judd), are large masses of hard-cemented greensand-rock, whose lithological characters are similar to the various boulders we have been describing. And we have seen that they contain some of the same fossil species, I conclude therefore (1) that the various blocks of greenish grit with *Terebratula ovoides* are of the same age as the derived fragments of dark grit in the Nodule bed at Upware, Potton, and Hunstanton; and (2) that they are all of Neocomian age, having been derived from a deposit closely connected with that of the Lower Neocomian sands of Lincolnshire, whose loose sandy materials being removed, the harder masses were left; some of them, perhaps, like 'sarsen stones,' never having been far removed from the place of their original construction, while others have been carried as 'drift' to greater distances. I believe, then, that the original home of *Terebratula ovoides* and *T. ovoides* var. *rex* was in the Neocomian seas, and not in the upper oolite.

GENERAL CONCLUSIONS AS TO THE 'DERIVED' NEOCOMIAN FOSSILS.

The presence of these 'derived' Neocomian fossils has had curiously different influences upon different observers, according to the different aspects in which they have been regarded. The Rev. T. Wiltshire, looking upon them as indigenous fossils, came to the conclusion that the Hunstanton carstone and ironsands were of the age of the Atherfield clay; Mr Teall, seeing that they were derived fossils, was, it would seem, thereby influenced to consider the Upware and Potton beds of later date than the horizon to which I believe they truly belong; whilst Mr Godwin Austen and Mr Seeley, recognising them as Neocomian species, and believing them to be indigenous, and seeing also their close correspondence in appearance and preservation to the Jurassic species (*Am. biplex*, &c.), concluded that these latter likewise were "denizens of the old sea-bed." As proceeding from this belief we have their theories of the cretaceo-oolitic age of the Farringdon and Potton series.

In the subjoined table the principal species of the derived
Neocomian fossils are given, together with an account of their
distribution in space and time:—

Ancyloceras gigas, Sby.   Atherfield clay ; Aptien of Bedoule, Bouches
de Rhône (d'Orb.).
Ammonites Cornuelianus, d'Orb.   Aptien of Paris area ; Upware,
Neocomian; Atherfield?; Hythe, Perte du Rhône, Aptien; Ber-
nese Alps.
      " A characteristic Aptien species," Pictet and Renevier.
Ammonites Martini, d'Orb.   Aptien, Paris area ; Atherfield clay,
beds 12—29, Fitton (Forbes and Ibbetson); Constantine, Africa,
Upper Neocomian ; Beausset, Var. Up. Neocomian.
Ammonites Deshayesii, Leym.   Atherfield clay, beds 6—10, Fitton ;
Aptien, Paris area ; Speeton, Upper Neocomian ; Brunswick,
Up. Neoc. ; Folkestone, Junction Bed ; Constantine, Africa ;
Beausset, Var. Up. Neoc.; Pyrenees, urgo-Aptien.
Perna Mulletti, Desh.   Neocomian proper of Paris area ; in England
mostly in the Upper Neocomian, Atherfield clay, and Cement
beds of Speeton ; Tealby, Mid. Neocom.; Schoeppenstedt, Lower
Neocomian ; Landeron, Lower Neocom. ; Switzerland, Valengien,
Urgouien.
Perna Ricordiana, d'Orb.   Neocomian proper of Paris area; Ather-
field beds, Shanklin sands, St Croix, Perte du Rhône, Aptien.
Thetis minor, Sby.   Shanklin, Speeton, Upper Neocomian.
Cardium subhillanum, Leym.   Upware, Neocomian ; Neocom. proper
of Paris area; Switzerland, Valengien ; d'Arzier, Landeron,
Marnes d'Hauterive, Lower Neocomian.
Trigonia ornata, d'Orb.   Hythe ; Atherfield Perna bed ; Perte du
Rhône, Aptien.

From this list we see that some of the derived species are
actually of late Neocomian or even Aptien age. *But the indigenous fossils prove the Nodule beds themselves to be of Aptien age.*
The same kind of anomaly seems to exist in the Neocomian
rocks of Mont Salève, where the fossils are casts of Neocomian
shells, &c., and the *Serpula* and Polyzoa have attached themselves
to the surfaces of these casts—a circumstance which M. de Loriol
states he is unable to explain[1].

Here then we are in a difficulty, for the history of a derived
fossil is usually a long and important one, identical indeed with
that of unconformity. We read its story as having lived its day
in the seas of its birth, and been buried in the accumulation of
sediment, when the surrounding rock penetrated it and it became

[1] See de Loriol, p. 5, "Mont Salève."

fossilized. Then by a series of changes which are commonly interpreted as first an upheaval and then a depression of the land, the rock formations were removed, so that the imbedded fossil shells were exposed and carried down to the shore, to be battered by the waves and otherwise injured by destroying agents, the whole process requiring a long period of time.

But just as a rock unconformity is often simulated by and may be confounded with *false bedding*, so also I believe the story of a derived shell may be greatly exaggerated by such an account as that given above. Many of the cases of derived fossils, including those we have under consideration, seem scarcely to admit of so prolonged a history as this; otherwise we should be obliged to refer our Midland Neocomians to at least as late an age as the gault, which is impossible.

I think we must admit that a very thorough fossilization may go on far more rapidly than is generally supposed. The rapid metamorphism of the interior of coral reefs is a well-known fact, and many other very modern deposits are quickly hardened into 'rock' especially in the presence of iron or carbonate of lime[1]. Rolled nodules of coal occur in the coal measures, and concretions are well formed in modern bogs. In the 'Challenger' dredgings concretions were found spread over the bottom of the deep sea. Just so, I believe, was the case with our Neocomian rocks, which contain derived Neocomian fossils. The ferruginous and phosphatic concretes were formed quickly after the formation of the rock, and, afterwards, some such alteration as a change of level or a variation in the strength or direction of the ocean currents, caused the sediments to be torn away again, and the older organisms were redistributed as 'derived' fossils in immediately subsequent deposits.

Looking to the geographical distribution of these derived fossils in the Neocomian rocks, we find that the Cephalopoda are essentially Southern species belonging to the Anglo-Gallic area, whereas we have found in the grit boulders evidence of a northern origin. This mixture is not surprising when we reflect that at this period the advancing southern sea had overwhelmed the Wealden estuary and spread itself northwards over the Eastern

---

[1] I have seen glacial moraine matter hardened into a strong breccia on mountain bog lands.

Counties where the two seas, northern and southern, gradually
approximating, at last surged together and fused into one continuous Upper Neocomian sea. [And this, be it noted, would be
almost sure to increase erosion, as there would be currents from
the one sea to the other. There is more than one way in which
depression means *erosion*. T. G. B.]

THE DERIVED FOSSILS OF THE WEALDEN.

I am not aware of any undoubted remains of Wealden fossils
having occurred in the Upware and Brickhill deposits, but in the
intermediate area, at Potton, rolled fragments of *Endogenites* have
been found in considerable numbers; also certain fragments of
sandstone in the collection of Mr J. F. Walker are very similar to
the Lower Cretaceous Sands (Wealden probably) of Shotover Hill,
Oxford. So that, as pointed out some years ago by Professor
Morris, it is probable that the Wealden rocks once formed part of
the shore line of the Neocomian sea beyond their present limits up
to the north of Brickhill; but we have no reason to believe they
ever extended so far as Cambridge.

It is the opinion of some geologists that the remains of Land
Vertebrates at Upware and Potton were derived from the denudation of the weald. This theory I cannot support, for some of
the specimens are beautifully preserved, being scarcely worn at
all; and the bones are not highly phosphatised, but are commonly impregnated with iron oxide. Moreover, the shore line and
land were close by, and we know that the Iguanodonts continued
to live on during Lower Greensand times. Therefore I regard
the great majority of these reptile remains of Potton and Upware
as having belonged to animals which lived and died during the
period of the Iron Sands and Coprolite Nodule beds. Some of
the Saurians and Crocodilians are exceptions to this rule, having
been washed out of the Kimmeridge Clay, and there are many of
them distinguished by being in a more highly phosphatised condition, with, occasionally, adherent lumps of phosphatic matter.

THE DERIVED FOSSILS OF THE PORTLANDIAN.

At Swindon the Portlandian rocks appear under two very different aspects. Below, we find the true marine Portlandians, but at the top we meet with 'dirt beds' and freshwater limestone interstratified with the marine limestones. For this latter set of beds I use the name Swindon Series[1].

A. *The Swindon Series.*

Remanie materials from the destruction of the Swindon series are not conspicuous either from the Upware or Potton Neocomian deposits, and they have not hitherto been recorded; but at Brick-hill undoubted fragments of the Upper Swindon Limestone are of frequent occurrence. Anyone who has worked in the great

[1] The section exposed in the Swindon Quarry in 1875 was as follows:

SWINDON SERIES.
- *h.* Greensand, Neocomian?
- *g.* Limestone bed with casts of Portlandian fossils. It varies rapidly in thickness.
- *f.* Limestone with profusion of fossils, moulds and casts, of Portlandian species, *Cytheræa rugosa*, *Cardium dissimile*, *Cerithium Portlandicum*, &c.
- *e.* Thin clayey bed, much contorted in the quarry.
- *d.* Bed of cherty rock and band of purer chert with freshwater fossils of Purbeck species [?] *Valvata?* Varies much in thickness. This bed is identical in appearance with some of the true Purbeck series.
- *c.* The dirt bed.

.... Unconformity ....

TRUE PORTLANDIAN.
- *b.* Great series of sand rock, and some sand; in places false bedded. Fossils few (*Trigonia, Ostrea*).
- *a.* Hard limestone, made up of a great extent of the casts of shells.

The beds *a* and *b* are thoroughly Portlandian, but in beds *c* to *g* we find some beds of Purbeck type. The chert bed and cherty limestone at *d* are just like those of the true Purbeck series, and one would unhesitatingly so call them were they not overlain by the limestone *f* and *g*, which are again Portlandian. But the latter beds are of greater importance in geological classification, being, in the language of Mr Blake, the more *normal* type, whereas the freshwater beds are truly *episodal*. The marine limestones *f* and *g* are not like the marine beds of the Purbeck series, but are truly Portlandian. And what we have here is the record of the gradually diminishing but oscillating stages of the latest Jurassic sea, when land and freshwater conditions resulting in formations of the Purbeck types were produced in Portlandian times. To this latter set of beds, ranging from *c* to *g* inclusive, I apply the term *Swindon series*. If we could trace this series southwards we should doubtless find it passing up into the true Purbecks, which were laid down when the boundaries of the old Jurassic sea had become much more limited.

Swindon Quarry will readily recognise the same rock and fossils (though now somewhat changed by being impregnated with phosphate of lime) among the Brickhill pebbles. Its specially characteristic fossil is the *Astarte cuneata* of Sowerby, and with these, associated in the same blocks, occur *Litorina* sp., *Trigonia* (small), *Lucina* sp., *Myacites* (very inequilateral sp.), *Natica, Lithodomus*, and *Ammonites biplex;* the latter with the pearly lustre of its shell well preserved.

Several of the Portlandian species mentioned below (e.g. *Cardium dissimile, Trigonia gibbosa* and *Lucina Portlandica*) may also have been derived from the same bed, for they are all found together in the Limestone of the Swindon series.

The rock as now found at Brickhill is a rather pale chocolate-coloured phosphate, with many coarse grains of sand.

To the freshwater beds of this same series I would refer the many fragments of silicified wood found at Potton and Brickhill.

From Upware the evidence of this series is less satisfactory. Some much rolled and worn casts of bivalves may pertain to the *Astarte cuneata*, but this is very doubtful, and I cannot consider that we have any conclusive evidence of the former existence of any part of the infra-cretaceous fluviomarine series, either Portlandian, Purbeck, or Wealden in this area.

B. *The Portlandian Fossils.*

The Portlandian species that have been recognised (some of them doubtfully) at Brickhill are the following: *Ammonites giganteus* (Sby.); *Ammonites biplex* (Sby.); *Myacites; Modiola,* sp.; *Cyprina; Cardium dissimile* (Sby.) ; *Myoconcha Portlandica* (Blake) ; *Trigonia gibbosa* (Sby.) ; *Trigonia incurva* (Sby.); *Pholadomya tumida* (Ag.)

Buccinum naticoides and *Lucina Portlandica* occur, I believe, at Upware.

Mr Walker has pointed out to me that many of these fossils (*Cardium dissimile*, &c.) occur in a very gritty phosphate, but they cannot be easily separated off from the other derived fossils by lithological characters.

I am inclined to think that a large proportion of the derived species of the Upware and Brickhill Neocomians were washed out of old coast lines made up of these uppermost Jurassic rocks.

The absence of *Cerithium Portlandicum*, the "Portland Screw," is noteworthy, but may be accounted for by the fragile nature of the casts of these shells.

KIMMERIDGE CLAY FOSSILS.

Kimmeridge clay fossils are amongst the most abundant of our derived shells. The following have been recognised:—

Belemnites explanatus, Phill.; *Ammonites biplex*, Sby.; *Ammonites Kœnigi*, Sby.; *Pleurotomaria reticulata*, Sby.; *Trochus*, sp. and other Gasteropoda (*Alaria, Chemnitzia*, &c.); *Gervillia aviculoides*, Sby.; ? *Lima læviuscula*, Sby.; *Cardium striatulum*, Sby.; *Myoconcha Sœmanni*; *Astarte Hartwelliensis*, Sby.; *Myacites recurva*, Phill. ?; *Myacites*, sp.; *Serpula tricarinata* (affixed to the Belemnites).

These are all (except the Belemnites) internal casts in a phosphatized condition.

Besides these invertebrate species the Kimmeridge clay seems to have furnished, by its destruction, some of the vertebrate remains now found in our Neocomians. I have no doubt that many of the Saurian remains are of Kimmeridge clay origin, amongst which some of the Ichthyosaurs, Plesiosaurs, and Pliosaurs are the most certain. Our 'Potton sand' specimens compare well with Jurassic species, and we have, besides, other positive evidence of Kimmeridge clay *remanié* material amongst the Molluscan remains. The Upware bed has furnished us with the teeth of *Pliosaurus brachydeirus*, both the three-sided and rounded forms, precisely like those of the Ely clay-pit (Kimmeridge clay); the Dakosaur teeth are likewise identical, and some of the Plesiosaurian, Ichthyosaurian, and Crocodilian teeth have probably the same origin.

It will however be remembered that very many of the reptiles of our Ironsand and Phosphatic beds were living in the Neocomian period, and are true natives of the deposit.

Amongst the Fishes are some extremely well-preserved teeth and jaws belonging to *Sphærodus, Gyrodus, Pycnodus, Strophodus*, and the spines of *Asteracanthus*, which are absolutely undistinguishable from the Jurassic species. Many of these are, in my opinion, 'derived' fossils of Jurassic age, but I fail fully to separate out these from others which certainly swam in the Neocomian sea.

Besides these there are *Hybodus* (dorsal spines, teeth, and *Sphenonchus*), *Otodus, Acrodus* and *Chimæroids*, all of which would be inscribed by Mr Walker amongst the 'derived' fossils.

CORAL RAG.

At Upware the Coprolite bed rests, in places, upon the Coral Rag, which was broken up by the waves so as to form a rubbly zone of fragments along the line of junction (see the section, p. 4) so that the origin of the following species found in the Upware Neocomians is not far to seek.

Chemnitzia Heddingtonensis, Sby.; *Cerithium muricatum* Sby.; *Gryphea dilatata*, var.; *Ostrea gregaria; Exogyra; Pecten vimineus; Lithodomus inclusus; Opis corallina; Unicardium; Hemicidaris intermedia* (flattened form); *Pseudodiadema hemisphericum; Cidaris florigemma ; Echinobrissus scutatus; Holectypus depressus; Glypticus hieroglyphicus;* ? *Apiocrinus* or *Millericrinus*.

Now it is a remarkable fact that these fossils have not become phosphatised but retain their original condition as seen in the Coral Rag pits near by.

But between Upware and Brickhill the Coral Rag is not developed as such, being represented by some coeval argillaceous clay deposits (the Ampthill clay of Prof. Seeley), and it is not unlikely that some of the species such as *Gervillia aviculoides* and *Lima læviuscula*, which we have placed in other lists, may have been derived from rocks of this age.

THE DERIVED FOSSILS FROM THE OXFORD CLAY.

This formation must have formed a considerable part of the shore line of the old Neocomian sea, for in many places the Ironsand and Nodule beds rest upon it. The *Ammonites* of this age as found at Upware and Brickhill are characterised by being preserved in oxide of iron (Limonite) and never phosphatised as those of the Kimmeridge clay and others are.

The Oxfordian species include:

Ammonites biplex (d'Orb. *non* Sby.); *Arca concinna ; Pholadomya Phillipsii ; Modiola bipartita; Rhynchonella varians*, var. *socialis.*

There was no Oxford clay exposed within some few miles of Upware, but yet the little Ammonites there are perfectly well preserved.

General. From the presence of these *remaniè* materials of Oxfordian to Neocomian ages inclusive and from the nature of the rock, it is clear that the Upware and Brickhill deposits were formed along the neighbourhood of a shore whose coast consisted of the upper and middle Jurassic rocks. In pre-cretaceous times the Kimmeridgian and probably the Portlandian rocks extended from Dorsetshire to Yorkshire, and the Purbeck and Wealden *conditions* also had a greater spread than now, probably reaching as far as Potton if not to Upware. The old coast line ran from S.W. to N.E. along a line to the west of Calne, Warminster, Farringdon, Oxford, Potton and Cambridge to our present Wash, and thence trending N.W. ran up into Yorkshire.

The following lists include all the derived species known to me from Upware and Brickhill.

I. The 'Derived' Fossils of the Upware Deposit.

Belemnites, Phragmicoe of
     ,,   *abbreviatus,* Miller.
     ,,   *explanatus,* Phill.
Nautilus, sp.
Ammonites biplex, Sby.
     ,,   *cordatus,* var. *Mariæ,* d'Orb.
Ammonites biplex, d'Orb., of the Oxford clay.
Ammonites Deshayesii, Leym.
     ,,   small smooth species (? Neocomian).
Ammonites Kœnigi, Sby. (collection J. F. Walker, M.A.).
Ancyloceras, sp. (J. F. W.).
     ,,   sp. in gritty phosphate.
Hamites.
Chemnitzia Heddingtonensis, Sby.
Cerithium muricatum, Sby.
Litorina (in dark grit).
Cerithium (ditto).
Pleurotomaria reticulata, Sby.
     ,,   sp.
? *Buccinum naticoides,* Sby.
Natica, 2 sp.

Nerinæa.
Aporrhais.
Turritella or *Chemnitzia.*
Exogyra (Coral Rag).
Gryphea dilatata, Sby. var. Coral Rag, &c.
Ostrea gregaria, Sby., Coral Rag.
Lithodomus inclusus, Phill., Coral Rag.
Modiola, sp.
Perna Mulletti (in dark grit).
     ,,   *Ricordiana,* d'Orb.
Pecten vimineus, Sby., Coral Rag.
Unicardium, Coral Rag.
Opis Corallina, Coral Rag.
Pectunculus, ribbed species, in phosphate.
Trigonia incurva, Sby. ?
     ,,   *gibbosa,* Sby. ?
Cardium striatulum, Sby.
     ,,   *dissimile,* Sby.
Lucina Portlandica, Sby.
Cyprina, large species.
Myacites, of Portland oolite, and Kimmeridge clay.

Cyrena rugosa, Sby.
Arca, or *Macrodon*.
Pholadidea.
Astarte ? (a Portlandian species).
   „ *cuneata*, Sby.
Terebratula ovoides, Sby.
   „   „   var. *rex*, Lan-
kester.
Serpula tricarinata, Sby. (attached
to *Belemnites explanatus*, Phill.).
Column of Crinoid (*Apiocrinus* ?).
Millericrinus ?
Hemicidaris intermedia, Flem.
Pseudodiadema hemisphericum Ag.
Cidaris florigemma, Phill.

Echinobrissus scutatus, Gmel.
Holectypus depressus, Lam.
Glypticus hieroglyphicus, Goldf.

Vertebrates :—

Sphærodus gigas, Ag.
Pycnodus.
Gyrodus.
Strophodus.
Hybodus.
Asteracanthus.
Plesiosaurus, several species.
Pliosaurus, several species.
Ichthyosaurus.
Dakosaurus.

II. THE 'DERIVED' FOSSILS OF BRICKHILL.

Those marked with a star (*) are not known to me from Upware.

Belemnites.
Ammonites biplex, Sby.
   „   „   „ the Ox-
ford Clay variety.
Ammonites giganteus, Sby. ?
   „   *Lamberti*, Sby.
Turbo, sp. (from the Kimmeridge
clay).
Pleurotomaria reticulata, Sby.
   „   spp.
Natica.
Chemnitzia.
Aporrhais.
Lima.
   „   *læviuscula*, Sby. ?
Gervillia aviculoides, Sby.
Arca, sp.
   „ *concinna*, Sby.
Astarte Hartwelliensis, Sby.
   „ *cuneata*, Sby. (of the
Swindon series).
Astarte, sp.
Myacites recurva, Phill.
   „   sp.
Modiola (Portlandian, sp.).
*   „ *bipartita*, Sby.
Cytheræa.
Thracia depressa, Sby.
Cardium striatulum, Sby.
Lucina.
Myacites.

Myoconcha Sœmanni.
   „   *Portlandica* ? Blake.
Pholadomya Phillipsii, Morris.
   „   *ovalis*, Sby.
Cyrena (*Cytheræa*) *rugosa*, Sby.
of the Purbeck Oolite.
Trigonia gibbosa, Sby.
   „   *incurva*, Sby.
   „   sp.
Rhynchonella varians, var. *sociale*,
Dav., of the Jurassic clays.
Rhynchonella, 2 sp.
Terebratula ovoides, Sby.

Vertebrates, Fish :—

Asteracanthus.
Sphenonchus.
Pycnodus.
Sphærodus gigas, Ag.
Edaphodon.
Otodus ? vertebræ of.

Reptiles :—

Dakosaurus.
Ichthyosaurus.
Pliosaurus.
Plesiosaurus.

Fossil wood :—

Silicified.

CHAPTER IV.

THE RELATIONS OF THE UPWARE AND BRICKHILL DEPOSITS TO OTHER BRITISH FORMATIONS.

OUR first care in considering the relations of the Upware and Brickhill deposits to other British rocks, is to establish the close relationship of these two deposits to one another. This is no difficult task. Homotaxially the beds are similarly placed, and there are many points of rock resemblance between them, amongst which the identity of the included phosphatic nodules is the most striking. But it is of the fossils that we have particularly to inquire, and these establish with decision the same fact of the close affinity of the two deposits. Out of the 86 species found at Brickhill, 61 are common to the Upware Bed. The numerous types of Brachiopoda, so special and limited in their distribution, which are common to both areas, are of themselves a sufficient witness of the nearness in age of the two deposits, only three of the Upware species being absent from Brickhill. Of the Brickhill Polyzoa all but two are found at Upware; and the sponges tell the same story, the *Elasmostoma (Manon) macropora, Catagma porcatum,* and *Verticellites clavatus,* species so restricted in space and time, being common to the two. Amongst the other organic groups the resemblances are however not so striking.

The state of preservation of all the fossils in the two localities is so precisely similar that the most practised eye fails to separate them out with certainty from a mixed lot.

Between the Upware and Brickhill areas is another well known place, Potton in Bedfordshire, where the Neocomian Phosphatic nodules have been worked for some years. This locality has been well described by MM. Brodie[1], Seeley[2], and Walker[3] and its close connexion with the Upware bed has been recognised by all.

The indigenous fauna of Invertebrates at Potton is scanty, but, so far as it goes, is the same as that at Upware. There are 21 species in common, mostly (13) *Lamellibranchiata,* including *Cyprina Sedgwickii* and *Plicatula Carteroniana;* also there are seven species of Brachiopoda, all of them Upware types : but all these fossils were rare at Potton though so very abundant in the other two districts.

That the two faunas, Upware and Potton, do not agree even more precisely with one another is doubtless due in great part to the different physical conditions which obtained in the two areas. Potton was no doubt nearer to the shore line of the ancient Neocomian sea than were Upware and Brickhill, and therefore the land animals are much more abundant in the former place. Also a further difference was due to the near presence of a great calcareous mass (the Coral Rag) at Upware, which was wanting at Potton. A number of important differences have resulted from this fact of the abundance of carbonate of lime at Upware, for this served not only to favour the growth of animals with calcareous supporting structures, but also furnished excellent matter for their petrification and permanent preservation. At Potton, where the lime was scanty, shell life was restricted, and a further result is, that even of such calcareous exuviæ as were buried in the bare sands, many have doubtless since been destroyed by the free percolation of water. Such indigenous species as do occur at Potton are now preserved in oxide of iron (Limonite), a mineral which does not readily serve for the mineralization of fossil remains. We have thus sufficient evidence of the very close relationship of the Upware, Potton, and Brickhill beds.

Beyond Brickhill, passing southwards by Oxford, we next meet with a number of isolated sand-hills which yield only the most scanty traces of fossils; and the first place that arrests us by its excellent sections and rich fossil remains is Farringdon, in Berkshire.

The Farringdon sponge bed is a well-known collecting ground and an old battle-field of geologists; it has been well described

[1] *Geological Magazine,* Vol. III. page 153.
[2] *Annals and Magazine of Natural History,* 1866, 1867.
[3] *Ibid.*

by Dr Fitton[1] in 1836, by Mr Godwin Austen[2] in 1850, and by Mr Sharpe[3] in 1854.

And here again the general relations of the bed are the same as those we have been describing in other districts, the iron sands resting upon the Kimmeridge Clay and Coral Rag, and being overlaid by the Gault, with marked unconformity at both junction lines, above and below.

The nature of the rock (a false bedded iron sand, as much like the Suffolk Crag as any other British rock) shews that we are still near upon the old coast line, as at Potton and Brickhill, and the included pebbles are also of the same nature, namely, Quartzite, Vein Quartz, Lydian stone, Flinty altered slate rock, and Jasper with (according to Mr Godwin Austen) water-worn crystals of Felspar; thus shewing that the waves of the Upware and Farringdon Neocomian sea beat against one and the same old coast line. The Phosphate of Lime nodules or "coprolites" also occur at Farringdon, but only scantily, and are indistinguishable in appearance from those at Potton and Upware; but here, as at Brickhill, the pebbles and 'coprolites' are loosely scattered through the sands and not collected into a distinct nodule or pebble bed as they are at Upware and Potton, and Rushmoor Bottom, in Bedfordshire.

Again, the close relationship of all these deposits to one another is scarcely less well shewn in their included vertebrate remains and 'derived' fossils, than by their native invertebrate species. Here at Farringdon we meet with the same vertebrate fossils, namely :—

Iguanodon.	Otodus.
Megalosaurus.	Gyrodus.
Crocodilians (Goniopholis? Telcosaurus, and Dakosaurus).	Sphærodus.
	Pycnodus.
Pliosaurus.	Strophodus.
Plesiosaurus.	Hybodus.
Ichthyosaurus.	Asteracanthus and Ischyodus.

And the same derived invertebrate species, namely :

Ammonites biplex, Sby.	Natica.
„ mutabilis, Sby.	Chemnitzia Heddingtonensis, Sby.
„ excavatus, Sby.	Modiola.
„ sp.	Arca.
Belemnites, large sp.	Cardium striatulum, Sby.

[1] Strata below the Chalk.
[2] Quarterly Journal Geol. Soc. 1850, p. 454.
[3] Ibid. Vol. x. p. 176.

K.   4

Myacites recurva, Phill.   *Ostrea dilatata*, Sby.
Astarte or *Cyprina* (same as at   ,,   *deltoidea*, Sby.
  Upware).   *Glyphea*.
Cardium dissimile, Sby.   *Cidaris florigemma*, Phill.
Perna mytiloides, Lam. ?   *Pentacrinus*.
Trigonia incuvra, Sby.   *Isastræa oblonga*, Flem.
Thracia depressa, Sby.   *Thamnastræa arachnoides*, Park.
Pholadomya.

These are found in a condition just similar to the derived fossils of the Cambridgeshire and Bedfordshire Neocomian areas.

Turning now to the native species for more positive evidence, we find the result is in perfect harmony with our other conclusions.

Of our Upware and Brickhill species 45 (a very large proportion) are known to occur also in the Farringdon gravels. These consist of 12 Lamellibranchiates, 16 Brachiopods, 4 Polyzoa, 3 Echinodermata, 2 worms and 6 (or more) sponges, and they include a number of forms of very limited range, which are of particular value in detailed and exact comparisons, such as *Neithea ornithopus*, Kpng.; *Opis Neocomiensis*, d'Orb.; *Lima Farringdonensis*, Sharpe; *Terebratella Menardi*, Lam.; *Terebratula prælonga*, Sby. and *T. microtema*, Walker.

The identity of the Sponges and Polyzoa of Upware and Farringdon is also very striking, all the more common Upware species being of frequent occurrence at Farringdon likewise.

Still there are some prominent differences which give to each of these two faunas a well-marked character of its own when contrasted with one another. This is apparent in a glance at any collection from the two places: but much even of this apparent isolation in the characters of the Upware and Farringdon *facies* was broken down with the discovery of the new fossil area at Little Brickhill near Bletchley. This little village lies just about midway between Upware and Farringdon, and it was extremely pleasing to find that its Lower Greensand fauna was likewise of just a corresponding intermediate character. Its close connexion with the Upware bed we have already seen, *supra*, p. 47, and its affinity to the Farringdon sponge bed is especially manifest in the abundance of *Peltastes Wrightii*, and the presence of *Lima Farringdonensis, Terebratella Menardi, T. oblonga*, and *Terebratula Tornasensis*; also in its Polyzoa and Sponges.

Leaving Farringdon we trace the Ironsands southwards to the Ferruginous conglomerate of Calne in Wiltshire with its rare and interesting fossil *Requienia Lonsdalei*[1] (*Chama, Diceras*).

A richer fauna was gathered together in 1849—1850 by Mr Cunnington from a road cutting at Sceend[2] near Devizes, where the Kimmeridge Clay was exposed overlain by a pebble bed with Quartz pebbles and fossils, which formed the base of a series of yellowish, dark green, and brown sands and iron sandstones, also containing fossils, "the whole surmounted with a patch of yellow brashy clay of a few acres in extent." This is evidently an exposure of the same series of Neocomian sands and gravels as at Farringdon, and the fossils leave no room for doubt in the matter.

In the Woodwardian Museum, we find from Sceend:—

Exogyra.	*Terebratella Fittoni,* Meÿer.
Lima longa, Roener.	„ *oblonga,* Sby.
Terebratula Tornasensis, d'Archiac.	*Rhynchonella depressa,* Sby.
Terebratella Menardi, Lam.	

and Mr Cunnington also records amongst others,

Emarginula Neocomiensis, d'Orb.	*Rhynchonella latissima,* Sby.
Waldheimia tamarindus, Sby.	„ *Gibbsiana,* Sby.
Terebratula sella, Sby. ?	*Opis Neocomiensis,* d'Orb.

Other species in the British Museum are, *Lima Farringdonensis*, Sharpe; various small univalves and bivalves; *Sphærodus Neocomiensis*, and various undetermined Reptile Bones.

A few miles further south, towards Warminster, these western shore deposits of the Upper Neocomian sea are lost to view, being overlapped by the newer members of the Cretaceous system.

There can therefore scarcely be a reasonable doubt remaining that these western Neocomian Ironsands and coprolite beds above described are all the result of one and the same physical phenomenon, or continuous (and not greatly protracted) series of phenomena; so that for most purposes they should be treated all together as one group. "No one can reasonably doubt the identity of the iron sand and gravels of Devizes, Rowde, and Calne, with those of Farringdon," wrote Mr Godwin Austen, so long ago as June, 1850. The many local differences in the several sections are, of course,

---

[1] According to M. Barrois this is a distinct shell from the *Diceras Lonsdalei* of the Middle Neocomian of Southern Europe.

[2] *Quarterly Journal Geological Society,* 1850, page 453.

of little importance in such a series of coarse and littoral deposits as these where rapid change and great variety are the general rule.

But besides these differences our more detailed studies amongst the indigenous fossils shew that there also exist other differences, which I take to be of real significance, between the several faunas. This is perhaps most strikingly seen in the species of Mollusca. Contrasting the Upware and Farringdon species, we find that they have no Cephalopod and only one Gasteropod (*Pleurotomaria gigantea*) in common, and of the Lamellibranchs only five species are known from both localities. Amongst the Brachiopoda there is more affinity, 11 out of the 26 Upware species being found at Farringdon.

In the Echinodermata the distinction is also marked; and even amongst the sponges it is not easy to see why the Farringdon species *Peronella furcata, Elasmostoma pezisa, Catagma (Manon) Farringdonense* and others should not have flourished equally well at Upware had the two formations been contemporaneous.

Between Brickhill and Upware we have the same kind of differences though not so strongly marked; the greatest contrasts here being amongst the Echinodermata and Lamellibranchiata.

Now remembering the close correspondence in the Physical conditions of these several areas, their topographical nearness to one another, and the similarity of the types of life in each locality, these differences are, to my mind, greater than one would expect to find along such a coast line now-a-days.

Wherever in two separate beds fossil organisms of widely different types are found—say, only fishes in one bed and only sponges in another, we have but poor materials for chronological comparisons; but with allied forms in two deposits which were formed under similar conditions, we have trustworthy data of the most exact value to work upon. Now bearing these considerations in mind the differences amongst the faunas of our several Ironsand and 'coprolite' beds become of such importance as to demand a special interpretation.

In working with the Brachiopoda there will also be found some reasons for supposing that the Brickhill bed was of somewhat older date than the Upware rock, the former having apparently been the centre of dispersion for many of our special types of Neocomian Brachiopoda (*Terebratula, Waldheimia,* and *Terebratella*), which reached Cambridgeshire at a somewhat later date. The general Physical phenomena of the period, which have been treated of by Professor Judd and Mr Teall[1], point in the same direction, and shew us that in the period of the deposition of these ironsands, a *transgressive* wave of depression was slowly creeping northwards from the south, so that the shore beds of the south are of somewhat older date than those further to the north. At the same time I consider that the whole of this series of ironsands, pebble beds, and coprolite beds of the western Neocomian outcrop although not of exactly the same age, are so nearly synchronous that they may fairly and, perhaps, in general work most advantageously be treated as one bed.

Now turning northwards from Upware and Cambridge we can readily trace the Ironsand group in a series of exposures through the Fen country up to the edge of the wash in the Hunstanton cliffs; but for the details of the stratigraphy of this area we must refer to the published work of Mr J. G. Harris Teall in the Sedgwick Prize Essay for 1873.

Just beyond Upware, at Streatham, and at Ely, Haddenham, Cottenham and near Rampton isolated patches of the Lower Greensand have been left and are now more or less well exposed. The Streatham bed was at one time worked for 'coprolites' (see Teall's Essay, p. 24), when it yielded fossils all of the Upware type. The 'derived' vertebrate fossils (*Dakosaurus*, &c.) were in a particularly good state of preservation at this place. But at the other localities no nodule bed has been discovered and only the most meagre traces of organic remains have been found. From the Neocomian of Ely specimens of *Pleurotomaria* (ornamented species); *Pecten orbicularis*, Sow.; *Corbis* (probably the same as at Tealby); *Trigonia*, sp. (large), are to be seen in the Woodwardian Museum.

North of Ely a neck from the great gulf of sea and fen-level stretches across to the east hiding all the solid geology under its beds of peat, silt, and gravel. This extends for some fifteen miles across (N. and S.), when the Lower Cretaceous sands again appear at the surface around Downham Market in Norfolk. At this place some new features present themselves and good fossil evidences may be gathered.

[1] Sedgwick Prize Essay.

In all the more southern developments that we have been considering the junction of the Lower Greensand with the gault is found to be abrupt and unconformable; here, however, at Downham Market there appears to be a perfect passage and transition from the sands below to the overlying gault[1]. An account of this section will be found *ante* [p. 11].

The thoroughly distinct nature and age of this bed from the Potton and Upware nodule beds is not recognised in the work of Mr Teall (Sedgwick Prize Essay[2]), but on the other hand all these deposits seem to have been taken as of approximately the same age, and all are referred to the Folkestone series. And this grouping has, I believe, had an unfortunate and false influence in appearing to support a *Folkestone sand* age for the Upware and Potton nodule beds. For the Downham Market truly belongs, as I also doubt not, to that age; but I find, from the fossil evidence, that this bed is totally distinct from all the other worked coprolite beds of the Eastern Counties, and must be kept clearly separated off from them; and thus the evidence rebounds to shew that if the Downham bed is of Folkestone age then the others are not of that age, but belong to some different and older period—a conclusion quite in harmony with the direct evidence from their contained fossils as already worked out. *See supra.*

The list of Downham fossils given by Mr Teall (p. 21) is, as he tells us, a mixed lot—partly gault and partly from the underlying sands and coprolite bed. Separating these out we find:

I. GAULT SPECIES:

Ammonites interruptus, d'Orb.   *Inoceramus sulcatus*, Park.
    „   *splendens*, Sby.   „   *concentricus*, Park.
Hamites, 2 sp.   *Nucula pectinata*, Sby.
Belemnites attenuatus, Sby.

These occur, all except the *Belemnites*, in a pale phosphatic condition, such as is frequent in the gault, and quite different from the Neocomian types.

---

[1] Mr Teall describes (in his Sedgwick Prize Essay, page 24) a gradual transition from the Ironsand to the gault at Gamlingay, Cambridgeshire. This section is now not visible.

[2] Mr Teall writes me that he has always believed the position of the Nodule bed varies in different localities.

II. NEOCOMIAN SPECIES:

Ammonites Beudanti, Brong.	*Janira Morrisii.*
Pleurotomaria.	*Spondylus.*
Solarium.	*Cyprina Liguriensis.*
Aporrhais.	„ *sp.*
Pecten orbicularis, Sby.	

This revised list does not however alter the value of the other facts already brought forward by Mr Teall, but only strengthens his conclusion that the Downham coprolite bed belongs to the *mammillaris* zone of the southern Neocomians.

Next, if we compare this bed with the Upware deposit we find that the differences between the two are very numerous and striking, whilst their resemblances are few. Both are phosphatic nodule beds with yellowish sandy matrix, it is true, but all the nodules at Upware are 'red coprolites'—the pale yellowish chocolate-like type, whereas at Downham Market there are, besides those of the gault, two types both equally and strikingly different from those of Upware (see *ante*), and indeed much more like those of the Cambridge greensand. It is remarkable that they are actually undistinguishable from those of the *mammillaris zone* at Grandprè in the Ardennes.

Now when we remember the great constancy in the characters of the Neocomian nodules all along the western Neocomian outcrop, and how great is their similarity in the other beds further north, we shall probably be prepared to attach considerable importance to the different character of the Downham coprolites.

Lastly, the fossil evidences conclusively decide the matter, for although the lithological characters of the two deposits are similar, and the general physical conditions must have been so nearly the same, the types of life being also similar, yet the only species common to the two beds are *Pecten orbicularis* and *Janira Morrisii*. Both of these are long-ranged types of no value whatever in detailed chronological comparison.

We must therefore reject this bed from our Ironsand series, as belonging to quite a distinct age, and as I am inclined to think, pertaining to quite another series of events. For I look upon the Ironsand group as belonging to the true Lower Greensand (Sandgate and Hythe series and perhaps base of Folkestone series), while the Downham Market coprolite bed might, it seems to me, better be taken as the basement bed of the gault;—important physical events having, as a rule, separated the two groups.

The Downham coprolite bed being thus excluded we are met with the demand—what then *is* the representative of the Upware Nodule bed in the Downham Market section ? I should consider that it very probably exists somewhere amongst the 20 feet of sand beneath the rock bed or Carstone, or, may be, a little deeper. These sands have not yielded any fossils, nor have I seen here any evidence of a physical break between them and the gault, other than that which is probably indicated by the Phosphatic nodule bed itself.

Passing northwards from Downham Market the Neocomian sands swell out considerably in thickness and form the sandy country about Sandringham warren. They vary somewhat in character, being sometimes a pure white sand, more often yellowish or fox coloured and frequently false bedded and obliquely laminated; while at Wolverton these sands "closely resemble those of Woburn, Potton and Sandy" (Teall, Sedgwick Prize Essay, p. 19).

In this country too the Carstone, "an indurated ferruginous sandstone," notifies its better development in well-defined crests and escarpments; and beneath it a bed of clay comes in which is worked for bricks and tiles at Heacham and Sandringham.

The fossil remains in all these places are most scanty and meagre, and I can only reproduce the records already published by Mr Teall, who found at Lodge Hill, Snettisham, specimens of *Lucina* and *Pecten orbicularis*, and at Heacham, in the stiff dark blue clay, *Ammonites Deshayesii, Pecten orbicularis* and a *Trigonia*.

Dr Fitton found the following fossils in some 'ferruginous masses' in a brickfield near Ingoldsthorp, Norfolk, which he records as follows[1]:

Cinulia incrassata, Sby.; *Avicula; Panopœa plicata*, Sby.; *Rostellaria calcarata*, Sby.; *Turritella granulata*, Sby.; *Venus faba*, Sby.

We have now arrived at the well-known Ironsands and Carstone of Hunstanton, a set of ironsands and pebble beds with a zone of nodules near the base of the series. Fossils occur rather rarely in this bed, in the lines of nodules just referred to, and we

---

[1] Fitton, *Strata below the Chalk*, page 317.

have given the list of species elsewhere (see *ante*, p. 33). Amongst these are *Ammonites Martini, Ammonites Deshayesii* and *Ancyloceras gigas*, which are all good Upper Neocomian or Aptien species very characteristic of the Atherfield clays—to which beds Mr Wiltshire has consequently referred the Hunstanton series. But so far as I have seen or been able to learn all the Hunstanton species are 'derived' fossils, and not native to the stratum in which they lie. They are all either in the condition of rolled phosphatised casts, or are found in the hard rolled lumps of dark iron grit as recorded *ante*, so that these fossils instead of proving the bed to be of Atherfield clay age, really shew that it is of some age posterior to that period :—though it is probably not far removed from it.

Comparing the Hunstanton bed directly with the Upware Nodule bed we find some resemblances between the two lithologically, also in the contained phosphatic nodules and pebbles; and particularly in the species of derived Neocomian fossils. To these points I attach considerable importance, and I consider that, notwithstanding the absence of a true indigenous fauna (they may have once existed there and since been destroyed), we have good reason to consider the Hunstanton and Upware sands and pebble beds as belonging approximately to the same age.

We have thus been enabled by the study of a number of isolated sections and their fossils to establish the close relationship of the beds of variously coloured sands, coprolite beds and carstones which are exposed, with some interruption, along an outcrop running from the N.E. at the Wash to the S.W. in Wiltshire[1]— stretching from the neighbourhood of Wicken in Norfolk down to Wicken in Cambridgeshire (!). They are again well developed around Potton and Sandy in Bedfordshire, and are last seen near the villages of Pottern and Sandy in Wiltshire (!!).

It now becomes a matter of great interest to discover the relation of this set of deposits to the better known and typical districts of the S.E. and N.E. of England.

In the year 1850 Mr Godwin Austen published a masterly description of the Farringdon and Swindon areas and came to the

---

[1] The Lower Sands of Shotover Hill, with Freshwater fossils, and the Downham Market coprolite bed excluded.

conclusion that the Farringdon bed was of Precretaceous age, and
was possibly the representative of the Portlandians. Mr Sharpe,
at the other extreme, argued that they were of much later date,
later than any other British cretaceous rock. Again in 1867 Mr
Seeley came to conclusions about the Potton sands similar to
those of Mr Godwin Austen with the Farringdon beds; but both
these authors seem to have given most attention to the vertebrate
fossils and to have believed that all these remains, together with
the rolled phosphatised casts of invertebrate species, were native
fossils proper to the deposit in which they occur. This entirely
destroys the value of their arguments, for we have seen that most
of these are really 'derived' fossils belonging properly to older
formations; and when these are excluded the apparent conflict of
evidence disappears. For the indigenous species point uniformly,
not to a precretaceous or postcretaceous age, but to the higher
beds of the Lower Cretaceous series, *i.e.* the upper Neocomian or
Aptien. This has been very generally recognised during the last
few years. Mr Walker writes[1]: "The age of the bed [at Upware] is
the same as that of the deposits at Potton, Farringdon and
Godalming, *viz.* upper Neocomian."

Mr Teall refers all the Ironsand series and coprolite beds to
the very latest part of the Lower Greensand, posterior to the
Hythe and Sandgate series, and he believes and considers it of
importance that these beds pass up by simple gradation into the
gault, so that they might indeed be taken as the basement beds of
the gault.

Mr Meÿer and M. Barrois take them to be an extension of the
Godalming Pebble bed.

I come to the conclusion that they are the representatives of
the Sandgate and Hythe beds of the South of England, including
the pebble bed, some of our "coprolite beds" presenting a very
close correspondence with the latter deposit. The upper part of
the Folkestone series is represented by the Downham Market bed,
and not by our Ironsand and phosphatic group, *a physical break
separating the two series.*

---

[1] In the Monograph of the *Trigoniæ* by Dr Lycett, *Palæontographical Society*,
Vol. xxix., page 145.

In England the lower cretaceous beds are developed under two very different types, namely, the North British Type or Speetonian, and the South British or Vectian. In Surrey, Sussex and Kent the Vectian series is divided as follows:

C. LOWER GREENSAND.
  (Upper Neocomian).
  (Aptien).
B. WEALD CLAY.
A. HASTINGS SANDS.

4. Folkestone Beds.
3. Sandgate Beds.
2. Hythe Beds.
1. Atherfield Clay.

But this series, so well developed in the S.E. of England and in the Isle of Wight, becomes greatly altered before it appears again some 30 miles off along the coast in Swanage Bay, for in this place the upper division or Lower Greensand is only with difficulty recognisable and is greatly reduced in thickness, though according to Mr Meÿer all the zones are there present. Another interesting peculiarity of this section is the intercalation of some fresh—or brackish—water deposits in the midst of the Lower Greensand, shewing the temporary recurrence of the earlier wealden conditions in this area. Still further east we meet with the anomalous and interesting silicious sandstone of Blackdown—the Blackdown beds; and seeing that this rock contains a number of Lower Greensand species, and that it lies exactly in the line of outcrop of the Ironsand series, we might reasonably expect it to be nearly related to these latter. But such an idea is decidedly overthrown by a study of the fossils, for although both rocks are shallow water formations, rich in organic remains, yet I only know of two species common to the two deposits, namely, *Cyprina rostrata* and *Pectunculus sublœvis*.

The Ironsand series of Hunstanton-Farringdon, and the true Vectian type of the Lower Greensand approach nearest to one another in Berkshire and Surrey, and it is in these places that we find, as already indicated by Mr Teall, the closest affinity between the two types. By the careful work of Mr C. J. A. Meÿer, F.G.S.[1] the series of sands, sandstones, limestones and clays in the country around Godalming have been correlated with the complete series

---

[1] *Proceedings Geologists' Association*, read December, 1868.

of Lower Greensand rock groups as developed further east on the coast of the Weald area in Kent.

The Godalming section is as follows (Meÿer):

		FT.
3.	*Upper series* (Folkestone beds, Meÿer, = Folkestone beds, Sandgate beds and upper part of Hythe beds, Geol. Survey).—Sands, carstones, and shelly limestones with pebble beds at base ... ... about	50
2.	*Middle series* (Sandgate and Hythe beds, Meÿer = greater part of Hythe beds, H.M. Geol. Survey).— Ash-coloured sands and sandstones, Ferruginous sands and sandy clay ... ... about	250
1.	*Lower series* (Atherfield clay).—Argillaceous beds with concretions of shelly limestone towards the base, about	160

Total 450—500

Comparing these in their order with the Upware and Brickhill beds, we find (1) the Atherfield clay quite dissimilar lithologically, while its palæontological relationship is by no means intimate. Mr Meÿer (who has kindly supplied me with revised copies of his papers) records 120 species from these beds at East Shelford, amongst which the following also occur as indigenous fossils at Upware or Brickhill:

Terebratula sella, Sby.
Rhynchonella lata, d'Orb. } Not the Upware varieties.
Exogyra Tombeckiana, d'Orb.
Ostrea Couloni, d'Orb.
Pecten Robinaldinus, d'Orb.
   „   *orbicularis,* Sby.
Perna Mulleti, Desh.
Arca Carteroni, d'Orb.
Cardium subhillanum, Leym.

(2) In the next set of beds (the Sandgate and Hythe series of Mr Meÿer), fossils are much less numerous, but the list of species shews strikingly the resemblance of its fauna to that of Upware. Twenty-one species are recorded by Mr Meÿer (loc. cit. pp. 8, 9), of which the specific names of ten have been determined, and all these ten species occur in the Upware and Brickhill deposits, namely:

Ostrea macroptera, Sby.   *Waldheimia tamarindus,* Sby.
*Rhynchonella Cantabridgiensis,*Dav. *Exogyra Couloni,* d'Orb.
Terebratula Meÿeri, Walker?   „   *Tombeckiana,* d'Orb.
   „   *sella,* Sby.   *Plicatula Carteroniana,* d'Orb.
   „   *prælonga,* Sby.   *Belemnites pistilliformis,* Blainv.

A layer of Phosphatic nodules occurs about 50 feet from the top of this series (Meyer).

(3) The upper series is considered under three divisions, (a) the Pebble bed, (b) the Bargate stone, and (c) the Upper sands and Car-stone.   The Pebble beds are stated to rest unconformably upon the sands beneath, and this fact, together with the nature of the deposit,—a coarse sand with angular pebbles—seems to indicate that this bed denotes a period of earth-movements and marine disturbances; a point which has been insisted upon by Mr Meyer as of great importance. The upper limit of the bed is ill-defined, passing on in gradual transition to the Bargate stone above. Now here again, just as with the Hythe and Sandgate series, every one of the pebble-bed fossil species that has been named and identified occurs also in the Cambridgeshire deposit. These are 15 in number:

Pecten Raulinianus, d'Orb.	*Waldheimia Juddii*, Walker.
„ *orbicularis*, Sow.	„ *pseudojurensis*, Leym.
Exogyra Tombeckiana, d'Orb.	„ *Wanklyni*, Walker ?
Terebratulina striata, Wahl.	*Terebratula depressa*, Lam.
Terebratella oblonga, Sow.	„ *extensa*, Meyer.
„ *Fittoni*, Meyer.	„ *microtrema*, Walker.
„ *Menardi*, Lam.	„ *Tornasensis*, d'Arch. ?
Waldheimia tamarindus, Sow.	

But the resemblances between this deposit and our 'coprolite beds' do not end here, for their lithological similitude is also most close; we have the same Lydian stones, jasper, irony grit, and quartzite, and *the same phosphatic nodules;* and again, just as at Upware, the bed becomes in places hardened into a conglomerate, so as to be almost identical with the lower conglomeratic coprolite bed of the Upware section. Again, we have the same fragments of fishes and derived fossils as at Upware including the following species identified from the collection of Mr Meyer, F.G.S., and Mr J. F. Walker, M.A., &c.:—

Lepidotus, scales and teeth.	*Gyrodus.*
Otodus.	*Acrodus.*
Pycnodus.	*Plesiosaurus.*
Hybodus.	*Rostellaria.*
Ammonites Lamberti, Sby.	*Lucina.*
„ *cordatus*, Sby.	*Myacites.*
„ *biplex*, Sby.	*Pentacrinus, &c.*

(*b*) Next we come to the Bargate stone, and here we find this fast increasing resemblance of the beds to the Upware type is abruptly broken off. The Bargate series consists of sand and layers of concretionary limestone very scantily furnished with fossil remains, but these have been perseveringly collected by Mr Meÿer to the number of 34 species, 24 of which are fully named. Amongst these 16 species are common to our coprolite beds, namely:—

Verticillites anastomosans, Mant.	*Terebratella oblonga*, Sby.
Pecten orbicularis, Sby.	„ *Davidsoni*, Walker.
„ *Dutemplei*, d'Orb.	„ *Fittoni*, Meÿer.
„ *Rauliniana*, d'Orb.	„ *Menardi*, Lam.
Plicatula Carteroniana, d'Orb.	„ *trifida*, Meÿer.
Exogyra Tombeckiana, d'Orb.	*Waldheimia tamarindus*, Sby.
Terebratulina striata, Wahl.	„ *Juddii*, Walker.
Terebratula depressa, var. *chrysa-*	„ *Wanklyni*, Walker.
lis, Lam.	

It is to the Godalming pebble bed therefore, that we must look for the most exact representative of the Upware Phosphatic Nodule Bed.

Now the above lists of species require more than a mere addition, and simple calculation of their proportionate numbers in order to see their full meaning; and their detailed consideration will strongly emphasise the facts gathered in our first oversight. The list from the Atherfield clay beds consists of species of little value for the work of detailed correlation, consisting as it does for the most part of types with wide ranges in the Lower Cretaceous series. And the same remark applies, though less thoroughly, to the middle sand group; still, we have here some of the special Upware types such as *Terebratula Meyeri* and *Rhynchonella Upwarensis*, and the presence of *Terebratula sella* and *T. præ-longa* are to my mind important, for these species do not occur in the overlying pebble series, and Bargate stone.

With the pebble bed, the value of the fossil evidence is especially intensified as we take each species into consideration. There is also the same preponderant development of the Brachiopods, including several of the most typical Upware forms.

In the list from the Bargate stone the species *Terebratella Davidsoni* and *T. trifida* are of importance, and my conclusion from the whole is that at Upware and Brickhill we have the representatives of all the Godalming beds from the Upper part of the

Hythe beds, (Meÿer) to the Bargate stone inclusive, (the whole being contained in the Hythe beds of H.M. Geological Survey.)

Each particular horizon in this series has however not been recognised at Upware and Brickhill, perhaps because of the unfortunate mixture of the fossils in the course of the workings; but we still have some evidence in the large blocks preserved in the Woodwardian Museum which shews that *Terebratula sella* (var. *Upwarensis*) was especially abundant in the lower phosphate bed, just as at Godalming it occurs in the Hythe beds and lower clays only. Other species from this conglomerate are,

Terebratella Fittoni, Meÿer.	*Terebratula depressa*, Lam.
Terebratula Moutoniana, d'Arch.	„ *prælonga*, Sby.
„ *microtrema*, Walker.	*Pecten orbicularis*, Sby.,

and many of the Lamellibranchs and Gasteropodes.

Further comparisons with more distant localities of the Vectian area yield results in perfect conformity with our conclusions from the Godalming area. In the Isle of Wight, amid the rich profusion of fossils found in the Atherfield beds, only 15 species occur in our Neocomian phosphatic series, and of these three only, namely, *Lima Farringdonensis* and the two sea urchins *Pseudodiadema Fittoni* and *Peltastes Wrightii*, belong to our particular Ironsand types. The 16 species common to the Upware bed and the Shanklin sands present a much greater proportion with respect to the whole faunas and form a much more weighty evidence of affinity. They include the 4 Brachiopods *Terebratella oblonga*, *Waldheimia celtica*, *W. tamarindus* and *W. Wanklyni*.

Again in East Kent we have the Hythe beds and Folkestone series well exposed, the former shewing in its contained fossils important relationship to the Upware and Brickhill beds.

Fifteen species of the Hythe Invertebrate fossils are also common to Upware, amongst which are

Ammonites Cornuelianus.	*Terebratula Moutoniana.*
Ancyloceras Fittoni.	*Waldheimia tamarindus.*
Plicatula Carteroniana.	*Peltastes Wrightii.*

Thus we find that the affinity of the Ironsand and Phosphatic series is not with the *mammillaris* zone nor with any part of the Folkestone series (H. M. Geol. Survey), nor do they belong to the

Atherfield clay period; but we do find a very close relationship between these beds and the Sandgate and Hythe series as developed in Kent and Surrey, and it is to this part of the Lower Cretaceous series that I consider they belong. The phosphatic nodule beds are especially similar to the pebble bed of Mr Meÿer as developed both in the east and west of the Weald area.

In proceeding to trace the relations of our Neocomians to those of the northern areas we find their connexions seriously broken at the two sea gulfs of the Wash and Humber; for these places just coincide with points of radical changes in the rock beds. The general characters of the rocks are so widely different on opposite sides of these two barriers that the detailed relations of the several types to one another are far from obvious. North of the Humber we have the Speetonian type, consisting of one great mass of stiff clay. Between the Humber and the Wash is the Tealby type of Neocomian consisting of

(3)   Upper sand group,

(2)   Middle series of limestones, ironstones and clays
        = Tealby series (Judd),

(1)   Lower sand group,

which Professor Judd has shewn to be the equivalents of the Upper, Middle, and Lower Speetonian (Neocomian) respectively. But the relations of these beds to the Ironsand and Carstone series over the other side of the Wash have yet to be determined.

Scarcely any fossils were known from the lower sands of Tealby at the time when Mr Judd's papers[1] on the Lincolnshire Neocomians were written; but in the summer of 1875 when working in Lincolnshire with my father, we were fortunate enough to find a number of beautiful fossils, mostly from hardened sandstone masses in the neighbourhood of Donnington and Claxby. These include:

Ammonites Kœnigi, Sby.	*Ammonites multiplicatus*, Roem. ?
„ *mutabilis*, Sby.	„ *plicomphalus*, Sby.

Species of

Chemnitzia.	*Pileopsis* and *Natica.*
Phasianella.	*Inoceramus.*
Pleurotomaria.	*Pholadomya.*
Trochus.	*Thetis.*
Crepidula.	*Lucina.*

[1] *Geological Soc. Proceedings*, August, 1867.

Cytherœa.
Avicula.
Arca.
Astarte and *Myacites.*
Lima Tombeckiana, d'Orb.
Trigonia Robinaldina, d'Orb.
    „   (dædaloid form).
    „   *alæformis,* Sby. var.

Trigonia Keepingi, Lycett.
    „   *Tealbyensis,* Lycett.
    „   *ingens,* Lycett.
Cucullœa Donningtonensis, Kpng.
    „   *errans,* Kpng.
Cardium subhillanum, Leym.
Pecten orbicularis, Sby. var. *magnus,* Keeping.

The middle series, or Tealby series proper, is very rich in fossils, which include the following Upware species:

Belemnites pistilliformis, Blainv.
    „   *subquadratus,* Roem.
Chemnitzia (with spiral grooving at base).
Ostrea Couloni, d'Orb.
    „   *macroptera,* Sby.
Pecten orbicularis, Sby.

Avicula Cornueliana, d'Orb.
Lima longa, Roem.
Lima Tombeckiana, d'Orb.
Astarte, sp.
Terebratula Tornasensis, d'Arch.
    „   *depressa,* Lam.
Waldheimia Juddii, Walker.

Our *Waldheimia Woodwardi* is represented by *W. hippopus* in Lincolnshire.

This list forms but an unimportant and insignificant proportion of the Tealby species, and they convey the impression that these beds are somewhat less nearly allied to our Ironsand and Phosphatic series than the Atherfield clay.

Unfortunately I have not been able to obtain many indigenous fossils from the Upper Neocomian sands of Lincolnshire, but it is to these I look as being the probable representatives of the Potton, Upware, and Hunstanton beds.

In our account of the phosphatic nodule beds (*ante* p. 10) we have already described the phosphatic nodule beds which occur in the Lincolnshire area, both at the lower junction with the Jurassic Clays and at the upper junction with the Red Chalk. But in spite of the close resemblance of the nodules in these beds to those of the Cambridgeshire and Bedfordshire district, I cannot regard either of them as having been synchronous with the Upware nodule bed.

Twenty-four of the Upware and Brickhill species are common to the Speeton area, and of these five are Lower, nine Middle, and eleven Upper Neocomian. The most noteworthy of these species in the lower beds are *Belemnites pistilliformis, B. subquadratus* and *Exogyra Couloni;* in the middle series *Belemnites pistilliformis, B. subfusiformis* and *Ostrea macroptera;* and in the upper beds *Belemnites subfusiformis, Ammonites Deshayesii, Ostrea macroptera, Tere-*

K.   5

bratula sella and *Vermicularia Phillipsii.* So that it is evident that if our Upware beds have any representative at Speeton it must be somewhere in the Upper Neocomian Clays of Prof. Judd. Such I believe may well be the case, for although the fossils are not nearly so similar as those in the Godalming district of the South of England, we must remember that the physical conditions at the time were very different in the two places, Speeton being deep sea whilst Cambridgeshire was shallow water or dry land; and if the Upper Neocomian earth-subsidence was at all uniform over the East of England, the Speeton waters were at their deepest while the Upware beds were being laid down further south. "What in England is called Lower Greensand is partly a littoral deposit (Potton, Farringdon, Hunstanton), and may readily be admitted to have its pelagic equivalent in some part of the Speeton Clay where *Exogyra sinuata* occurs [1]."

[1] Phillips, *Geology of Yorkshire*, 2nd edition, page 99.

CHAPTER V.

THE FOREIGN RELATIONS OF THE UPWARE AND BRICKHILL NEOCOMIANS.

THE old sea-basin in which the Upper Neocomian rocks of the South of England were deposited extended far to the south towards the centre of France till it reached the granite plateau of the Auvergne and Puy de Dome: and in an east and west direction it was bounded by the ancient rocks of Normandy and of the axis of Artois (Ardennes region)[1].

But in the earlier part of the Neocomian period (Lower Neocomian) this sea was not quite so extensive as here indicated, barely reaching as far north as the present north coast of France where, at that time, the great Wealden river was laying down some of its immense accumulations of freshwater sediments. In this part of the series therefore it can scarcely be possible to make any detailed correlation between the South of England freshwater beds and their marine representatives in the French area.

But our Ironsand and Phosphatic series was laid down during the immediately subsequent period when this Neocomian sea had attained to its greatest extent and was united with the northern (Anglo-Germanic) sea, through what Mr Teall[2] calls the North-Eastern channel.

Now the deposits laid down in the deep water of the Upper Neocomian sea differ considerably from those formed around the margins; and in searching for the extension of our Cambridgeshire and Bedfordshire Lower Greensand rocks, these being shallow-

---

[1] A tongue or neck of land running across the eastern counties separated this old sea from the northern (Anglo-Germanic) ocean.

[2] Sedgwick Prize Essay, 1875, page 40.

water formations, we naturally look for their best marked representative beds to occur along other parts of the old shore line.   In
the Ardennes district these old coast deposits are again well developed, and in some places they appear actually to be extremely
similar to our Ironsand series.   They have been thoroughly worked
out by M. Barrois[1], who refers them principally to the *Mammillaris*
zone[2], and the Upper Neocomian or Aptien.

In the comparison of the Upware and Brickhill fossil species
with these beds we find that (1) they shew but little affinity to the
Mammillaris zone, just as we found in England.   The Upware bed
does not belong to this zone, but still there are some eight or ten
species common to the two deposits, including (see Barrois, p. 271)
Terebratula capillata, d'Arch.; *Terebratula Moutoniana*, d'Arch.;
and *Waldheimia Juddii*, Walker.

(2)   A far more important list of species (more than 20) links
the Upware bed with the Lower Aptien clays and sands of the
Ardennes.   These include *Pecten Dutemplei*, d'Orb.; *Spondylus
Roemeri*, Desh.; *Verticillites anastomosans*, Mantell; *Elasmostoma
acutimargo*, Roemer; *Elasmostoma pezisa*, Goldf.; *Peronella furcata*, Goldf.; *Terebratula depressa*, Lam.; *Terebratula Moutoniana*,
d'Arch.; *Waldheimia pseudojurensis*, Leym.; and *Waldheimia tamarindus* var. *magna*, Walker.

Several of these are very characteristic of our Ironsand and
Phosphatic series.   Moreover the lithological resemblances are
also, according to M. Barrois, very close, for that gentleman writes
(l. c., p. 257): "J'ai été frappé de l'analogie du Gompholite de
Blangy avec le *Farringdon Gravel;* l'aspect des couches est le
même, les caractères de leurs faunes sont identiques; il y a enfin
à Blangy des formes connues (*Verticillites anastomosans, Terebratella truncata, Cidaris Farringdonensis*) qui sont si caractéristiques
du gisement du Farringdon, que je crois que tous les géologues
qui compareront ces deux localités n'hésiteront pas à les identifier
malgré leur éloignement et leur isolement;" and I have no hesitation in adopting his views that the lower Aptien beds of the

[1] "Mémoire sur le terrain crétacé des Ardennes et des régions voisines, par le
Dr Charles Barrois," *Annales de la Société Géologique du Nord*, tome v., page 227.
Lille, 1877—78.

[2] By English geologists the *Mammillaris* zone is usually placed in the Folkestone
series as the uppermost bed of the Lower Greensand, but on the continent it is put
into the Gault.

Ardennes are truly the representatives, in part at least, of our Upware, Potton, and Farringdon sand and Pebble beds.

In the central area of the Paris Basin the precise horizon of our Phosphatic beds does not appear to be so clearly indicated. Much discussion has been carried on as to the age of the Atherfield beds with reference to the Parisian series; but all agree that the upper beds of the Vectian area, to which our phosphatic series pertains, are of Aptien age; and according to M. Barrois the Farringdon beds might be identified with the *Argile à plicatules* of the Haute Marne.

In Switzerland the Neocomian fossils have been admirably worked out in the magnificent monographs of MM. Pictet, de Loriol, Renevier, and Campiche, and we find a considerable number of our species are there described and figured. Thirty-one species are common to the Upware and Brickhill beds and the Lower Cretaceous of St Croix.

To the North of Geneva, and at Mont Salève, we have the Jurassian type of Lower Cretaceous which is divided into the

    (4)   Aptien or Upper Neocomian.

    (3)   Urgonien or Middle Neocomian.

    (2)   Neocomien or Lower Neocomian.

    (1)   Valengien.

The Valengien deposits are known with certainty only over a very limited area; and its fossils as developed at d'Arzier (Vaud), have been worked out by M. de Loriol[1]. Amongst them we find seven species of Upware fossils, including *Janira atava*, Roemer; *Opis Neocomiensis*, d'Orb.; *Waldheimia pseudojurensis*, Leym.

As examples of the Swiss Neocomian we may take the Mont Salève or Landeron sections. From the Marnes d'Hauterive (Lower Neocomian) of Landeron twelve native Upware species are recorded besides two of our derived Neocomian shells, namely :

[1] *Palæontologie Suisse*, Vol. IV.

MARNES D'HAUTÉRIVE, LANDERON[1], LOWER NEOCOMIAN.

Pycnodus Couloni, Ag.
Serpula antiquata, Sow.
[*Panopœa Neocomiensis*, d'Orb.]
Cardium subhillanum, Leym.
Opis Neocomiensis, d'Orb.
[*Perna Mulletti,* Desh.]
Lima Tombeckiana, d'Orb.

Janira atava, d'Orb.
Plicatula Carteroniana, d'Orb.
Ostrea Couloni, (Defr.), d'Orb.
   „   *Tombeckiana,* d'Orb.
   „   *rectangularis,* Roem.
Terebratula tamarindus, Sow.
   „   *sella,* Sow.

And from the Pierre jaune (Middle Neocomian) of Landeron, we also have 12 Upware and Brickhill species, these beds further resembling our phosphatic deposits in their rich sponge fauna.

UPWARE SPECIES IN THE PIERRE JAUNE.

Pycnodus Couloni, Ag.
Sphærodus Neocomiensis, Ag.
Lima Tombeckiana, d'Orb.
Pecten Robinaldinus, d'Orb.
Hinnites Leymerii, Desh.
Ostrea Couloni (Defr.), d'Orb.
   „   *macroptera,* Sby.

Ostrea rectangularis, Roem.
Terebratula sella, Sby.
   „   *Moutoniana,* d'Arch.
   „   *tamarindus,* Sby.
Elasmostoma acutimargo, Roemer.
Catagma cupuliformis, From.

The following list also shews considerable relationship between our Upware beds and the Swiss Aptien.

FOSSILS OF THE APTIEN BEDS, PERTE DU RHONE, ALSO OCCURRING IN THE IRONSAND SERIES.

Belemnites subfusiformis, d'Orb.
Cyprina angulata, Sby.
Janira Morrisii, P. and R.
Opis Neocomiensis, d'Orb.
Ostrea Couloni, d'Orb.
Exogyra conica, Sby.
Panopœa plicata, Sby.

Pecten Dutemplei, d'Orb.
Pinna Robinaldina, d'Orb.
Pleurotomaria gigantea, Sby.
Serpula antiquata, Sby.
   „   *gordialis,* Goldf.
Terebratella oblonga, Sby.
Rhynchonellæ and *Terebratulæ.*

Thus, from the comparison of our fossils with the various lists of Swiss species, we gain little knowledge of their exact affinity to any one particular zone in the Swiss series.

---

[1] De Loriol and Gilliéron, *Monographie de l'Étage Urgonien inférieur du Landeron* (Canton de Neuchâtel). Geneva, 1869.

South of Mont Salève, near Geneva, we meet with quite a different type of Neocomian rocks with a fauna almost entirely distinct from our British forms, and indeed very different from the north Swiss or Jurassian group. The great Diceras limestones and *Terebratula diphyes* limestones and other Neocomian limestones are enormous calcareous masses, making up in their simple thickness such great mountains as Mont Ventoux, 1910 mètres high, in Provence. These beds are hardly separable in places from the Jurassic limestones below or the Tertiaries above.

Although there are so few common species to connect this great Limestone series with our British Lower Greensands it is still interesting to note the very wide range of some few of them, such as *Ammonites Martini, A. Deshayesii,* and *Belemnites pistilliformis* which are recorded by M. Coquand from the province of Constantine in Africa; while as many as 13 of our Upware and Brickhill species are known (Magnan) from the Urgo-Aptien beds of the Pyrenees.

SPECIES COMMON TO THE IRONSAND SERIES AND THE NEOCOMIAN OF THE PYRENEES (MAGNAN).

Terebratula sella, Sby.;   *Terebratula Moutoniana,* d'Arch.

SPECIES COMMON TO THE IRONSAND SERIES AND THE URGO-APTIEN OF THE PYRENEES.

Belemnites semicanaliculatus, Bl.	*Ostrea carinata,* Sby.
Ammonites Deshayesii, Leym.	„ (*Exogyra*) *Tombeckiana,*
Pecten (Neithea) atavus, Roemer.	d'Orb.
Spondylus Roemeri, Desh. ?	*Terebratula sella,* Sby.
Hinnites Leymerii, Desh.	„ *praelonga,* Sby.
Ostrea Couloni, Defr.	„ *pseudojurensis,* Leym.
„ *macroptera,* Sby.	„ *Moutoniana,* d'Arch.

NORTH EUROPE.

In Germany the cretaceous rocks occupy but a very small area of the surface of the country. Sections or other exposures of the Neocomian rocks are few and far between, and the relations of the various rock series to one another are very difficult to determine, so that I am by no means satisfied that any of the various theories of correlations of these beds has been fairly established. The similarity of the sponge faunas of our phosphatic beds and Essen in Westphalia[1] deserves to be mentioned, although the rock of this latter place is generally referred to a much higher horizon, viz. the Tourtia or Cenomanian. Our *Peronella furcata* and *Elasmostoma pezixa* are I have no doubt the same as the original Essen species, and *E. acutimargo* and the curious *Pachytiloda* have also very near representatives amongst the Essen forms.

The great series of brownish yellow sandstones of the Teutoburger Wald, near Bielefeld, shews no special palæontological affinity to our Upper Neocomian (Lower Greensand) sands.

In the country around Hanover we again meet with a great Lower Cretaceous freshwater series like our Vectian type of Wealden; but the overlying clays shew little community either of lithological or palæontological character with our ironsand series; nor indeed with any part of the Vectian cretaceous group.

Still further east at Salzgitter, south of Brunswick, we meet with some open-air iron-workings in the Neocomian rocks which are placed by Professor Judd in his Middle Neocomian group. The similarity of these beds to the oolitic iron-beds of Lincolnshire has been pointed out by Mr Judd (*Quart. Journ. Geol. Soc.* 1870, p. 330), who writes, "The English and German ironstones are, both as rock specimens and in polished sections under the microscope, quite undistinguishable in their characters." Specimens of *Pecten cinctus* from the two areas are almost identical in every respect[2].

As compared with the Upware bed this rock is interesting as

[1] The fossils of this bed were found in the course of the workings for coal around Essen; these having now ceased the cretaceous fossils are difficult to obtain.

[2] Some of the iron ore is as perfectly oolitic in appearance as the well-known Pisolitic ironstone of Lincolnshire, the grains being round and regular, and specimens may be obtained identical with the Lincolnshire rock. But generally certain larger angular and sub-angular fragments occur scattered through the rock, and it frequently passes into a rock composed of a blue clay matrix, crowded with iron fragments, most of which are markedly angular. Now some of these latter are quite evidently fragmentary rocks, the iron particles being more or less worn fragments, of irregular shape. Again this type passes by perfect gradation into the most thoroughly rounded oolitic-looking rocks. As a result of my work in this quarry I became convinced that this rock *is* no true oolite, but is made up of the rounded and polished fragments of iron derived from older rocks; perhaps from the destruction of the Brown Jura formation. This is also the opinion of Herr Geheimer von Strombeck of Brunswick. To such rocks I apply the term '*pseudoolitic*.' An examination of thin sections confirms this opinion as a whole, but discloses also a number of true oolitic grains in most specimens.

containing large isolated phosphatised masses of calcareous matter, pale in colour, which are much tunnelled by Lithophagi (*Lithodomus* probably), and as having very similar-looking phosphatic derived fossils, such as *Ammonites Lamberti, Am. Conybeari, Myacites,* Ichthyosaur vertebræ and fossil wood. It is also remarkable for containing the same Lydian stones and quartz pebbles and smaller grains as at Upware in the coarser beds. A similar bed of ironstone, but much thinner, is found at Goslar at the north foot of the Harz mountains.

The Upper Neocomian beds (Judd) in all these places are but poorly exposed and have yielded few fossils, so that we have small means of comparison with our English series.

But it is in the Brunswick Neocomians at Schöppenstedt that we find the most interesting set of resemblances to our Upware and Brickhill phosphatic beds. In this country we have a series of clays above and limestones below, which have been divided into zones and their fossils collected by Herr Geheimer von Strombeck of Brunswick. The clay series is considered by Prof. Judd to represent his Upper and Middle Neocomians, and the limestone series he regards as Lower Neocomian. It is in the former group that we should have expected to find the representatives of our Upware beds. Such however is not the case, for in Prof. Judd's lists (*Quarterly Journ. Geol. Soc.*, Vol. XXVI.), *Ammonites Deshayesii,* Leym., and *Terebratula Moutoniana,* d'Arch., are the only species common to our phosphatic beds and the Brunswick Upper Neocomian. On the other hand *Avicula Cornueliana,* d'Orb.; *Panopæa plicata,* Sby.; *Lima longa,* Roemer; *Exogyra sinuata,* Sby.; *Terebratula Moutoniana,* d'Arch.; *Rhynchonella antidichotoma* and *Serpula Phillipsii,* Roemer; are common to the Middle Neocomian. We must however remember that much of the want of correspondence between the fossils of these beds and our Upware fauna is undoubtedly due to the difference in the prevailing physical conditions.

But coming now to examine the lower group of Limestones, &c., we find a remarkable set of repetitions of the characters of our Upware and Brickhill phosphatic series. These beds are exposed in several quarries near Berklingen to the S.W. of Schöppenstedt, and the following section is seen to the east of the village of Grosser Vahlberg:

(5)   Limestones similar to (1),

(4)   Second Pebble bed,

(3)   Limestone bed,

(2)   First Pebbly bed with iron fragments,

(1)   Beds of pale yellowish and cream-coloured Limestone,
the whole exposure amounting to 10 feet.

At the first glance this section does not strike one as being
very like that of Upware or Brickhill, for this is altogether a calcareous series, while the Upware section is in the main a sand series.
But the pebble beds (Nos. 2 and 4 of section) are astonishingly
similar to our Upware conglomerate of the lower coprolite seam.
The general appearance of the two is identical though some difference is found in the greater proportion of iron fragments in the
Schöppenstedt rock.   Quoting from my pocket-book I find we
have "The same angular iron-fragments; the same phosphatic
nodules, identical in appearance with those of Upware; the same
derived fossils, and in the same mineral condition ; and the same
angular fragments of Lydian stones."

Looking at a series of the indigenous fossils too, their resemblance to the Upware and Brickhill ones is even more striking,
and the aspect of a drawer of Schöppenstedt fossils is almost identical with our Upware series, the adherent rocks being absolutely
undistinguishable one from the other.   So that although one had
thought that our Neocomian phosphatic bed fossils had so peculiar
a character that they could at once be recognised amongst any
other specimens all over the world, I now find it necessary to take
special care to keep the several sets of Upware, Brickhill and
Schöppenstedt fossils separate.

Looking to the groups of life we find that these also are similar,
the same kinds of Sponges and Polyzoa, Brachiopods and Lamellibranchs being found in the two areas.

The following list of the Schöppenstedt fossils shews a close
affinity between these two deposits :

FOSSIL SPECIES FROM THE NEOCOMIAN ROCKS OF SCHOEP-PENSTEDT, BRUNSWICK.

Elasmostoma acutimargo, Roem.
Cidaris, plates, and thorny spines of.
Serpula gordialis, Goldf.
   ,,   *antiquata,* Sby.
   ,,   *quadrangularis,* Roem.
   ,,   *plexus,* Sby. ?
   ,,   *lophioda,* Goldf.
   ,,   *ampullacea,* Sby.
Vermicularia Phillipsii, Roem.
Terebratula prælonga, Sby. ?
   ,,   *microtrema,* Walker ?
   ,,   *sella,* Sby.
   ,,   ,,   var. *Tornasensis,*
   d'Arch.
Waldheimia celtica, Morris.
   ,,   *hippopus,* Roem.
   ,,   *Juddi,* Walker ?
   ,,   *tamarindus,* Sby.
Terebratella Pucheana, Roem.
   ,,   *oblonga,* Sby.
Rhynchonella, several species.
Ceriopora nodosa, Kpng.
   ,,   *polymorpha,* Goldf.

Entalophora dendroidea, Kpng.
Heteropora ramosa, Roem.
Ostrea Couloni, d'Orb.
   ,,   *frons,* Park.
Exogyra spiralis, Roem.
   ,,   *Tombeckiana,* d'Orb. ?
Avicula Cornueliana, d'Orb.
Pecten Robinaldinus, d'Orb.
   ,,   *Carteroniana,* d'Orb.
   ,,   *elongatus,* Lam.
   ,,   *cinctus,* Sby.
   ,,   *orbicularis,* Sby.
   ,,   *(Neithea) ornithopus,* Kpng.
Lima longa, Roem.
   ,,   *Tombeckiana,* d'Orb.
Plicatula Carteroniana, d'Orb.
Perna Mulleti, Desh.
Trigonia carinata, Ag. ?
Spondylus Roemeri, Desh.
Myacites ?
Lithodomus.
Patella.
Belemnites subquadratus, Roem.

PART II.
SPECIAL PALÆONTOLOGY.

I. VERTEBRATA.

REPTILIA.

THE Reptilian remains from Upware and Brickhill, although few in number, include representatives of most of the types which are so abundant and well known from the Potton area in Bedfordshire[1]. Remains of Deinosaurians, including *Iguanodon;* of Crocodilians, including *Goniopholis* and *Teleosaurus;* and of Saurians proper (*Plesiosaurus, Pliosaurus* and *Ichthyosaurus*) all occur, and the bones are usually in a good state of preservation, being much less wave-worn than the Potton specimens.

It is unnecessary here to describe in detail all these remains, for many of them are not true cretaceous species at all, being only the wave-spoils from the old sea cliffs of Jurassic rocks. Examples of these latter are seen in the teeth of *Pliosaurus brachydeirus,* and *Dakosaurus,* not uncommon fossils, which are identical with those found in the Kimmeridge clay of the immediate neighbourhood; also they are types which are not known to have lived on into the cretaceous epoch. These are therefore, with others, 'derived' fossils belonging properly to earlier geological periods.

But others of the Reptilian remains, such as the Deinosaurs, and the crocodilian teeth, are cretaceous in type and are just such

[1] See Seeley, *Index to Reptilia,* etc., *in the Woodwardian Museum,* Cambridge, 1869.

creatures as we should expect to have lived on the neighbouring land in the Lower Cretaceous period.

Many of the Saurians, especially the *Plesiosaurs*, were also, I believe, native around the Lower Greensand shores; their backbones are often excellently well preserved, and but little rolled; and although the projecting structures of their neural arches and the perforations for the nutritive arteries offer just such protected recesses as the phosphatic substance is generally formed in, yet no such phosphatic matter is found there, and the bone itself is often mineralized more with oxide of iron than phosphate of lime. See *ante*, p. 18.

It has been argued, and is still maintained by Prof. H. G. Seeley, that all the remains we find at Potton and Upware belong properly to the bed in which they are found, the animals having lived at the time these beds were deposited. Mr J. F. Walker, on the other hand, advocated the theory that all the Reptilian remains are of *derivative* origin, having been obtained by the destruction of the Jurassic and older Cretaceous rock groups, and this has been the generally received theory. But the truth is, I believe, as stated above, that both these ideas are in part correct, some of the species being 'derived' while others actually lived at the period.

One of the principal means of separating these two groups is found in an important difference of their lithological nature, namely the 'derived' fossils are found, for the most part, thoroughly fossilized in phosphate of lime, whereas the true Neocomian Vertebrate remains are preserved in oxide of iron. Moreover, as might be expected, the former group has suffered much more from abrasion by water-rolling than the latter[1]. Thus I have been enabled to separate out two different sets of vertebrate remains, the one of contemporaneous Lower Cretaceous age, and the other properly of Jurassic age.

Additional evidences of the truth of this conclusion are that certain types of Plesiosaurus vertebræ which occur in the Upware and Potton beds are not known in the Jurassic rocks, and on the other hand it is a striking fact that the large Saurian vertebræ,

---

[1] The collection of bones from Potton in the Woodwardian Museum cannot be taken as representing their usual state of preservation, since only the best specimens have been selected there.

mostly *Pliosaurus* and *Ichthyosaurus*, which are the most common
and conspicuous fossils of the Kimmeridge and Oxford clays, are
absent from the Upware and Brickhill deposits[1].

DEINOSAURIA.

Of the Deinosaurs two teeth of *Iguanodon* have been found at
Upware, now in the collection of Mr J. F. Walker, of York; and a
well-preserved caudal vertebra (*Pelorosaurus?*) is in the Woodwardian Museum (VII. $\frac{b}{5}$, Seeley, *Catalogue of Reptilia*, &c., Cambridge, p. 78). None have occurred at Brickhill.

The figure (Plate I. fig. 1) represents a small claw-like dermal
spine of a Deinosaur from Upware, also in Mr Walker's collection.

An Iguanodon tooth, with its serrated edges perfectly preserved, has been found by Mr C. J. A. Meÿer, F.G.S., in the
Bargate stone at Guildford; and the Cambridge University Museum
possesses a fine series of them, usually more or less worn, from
Potton.

SAURIANS.   *Plesiosaurus.*

Some of the Plesiosaurian remains from Upware are recorded
by Professor Seeley in his *Catalogue* (Reptilia, &c. p. 75 et seq.),
namely v. $\frac{9}{15}$, which he states "indicates a species different from
all the Potton forms"; and v. $\frac{b}{8}$, which is the cervical vertebra of " a
well marked species." The centrum is very transverse (length
1¼ in., breadth 1¾ in., depth 1 in.); the articular faces are shallow
(not deeply concave), with a slight central prominence, and the
neural arch is slender. v. $\frac{4}{8}$ and v. $\frac{4}{55}$ are dorsal vertebræ, the
former approaching the *P. Neocomiensis* of Campiche in form.
(See Pl. I. figs 2, 2a.) Other vertebræ are arranged in shelf *d*, and
specimens of the paddle-bones—*humerus* and *femur*—are in v. *e*.

Besides these there is a set of five vertebræ, mid-dorsal and
late dorsal in position, well characterized and in unusually good
preservation. The centrum is short-cylindrical, the sides and
articular faces nearly straight, but slightly concave. Length 35 mm.,
breadth 38 mm., depth 35 mm.

---

[1] These large vertebræ *are* found in the Potton bed, and it is important to ob-
serve that they occur there in a highly phosphatised condition and with masses of
phosphate adhering to them; quite different from the ordinary bones of the
Iguanodonts, &c.

From Brickhill also a few small vertebræ have been obtained, some of them belonging to the species just described, and others (Pl. I. figs. 3, 3a) with a very transverse short centrum, the sides and articular faces nearly flat, and the outline of the latter roughly trapezoidal. The neural arches are lost, their osseous union with the centrum having been, as in *Ichthyosaurus*, incomplete. The pair of pits on the ventral surface for the entrance of the nutritive arteries is conspicuous in all the specimens. Measurements : length 28 mm., breadth 48 mm., depth 39 mm.

Some slender, curved, pointed teeth from Upware probably belong to *Plesiosaurus*. Their surface is covered with small, very continuous primary ridges, upon the sides of which there usually exists a more delicate, crimpy crest. They measure 35×12 mm.

PLIOSAURUS.

Of the Pliosaurian teeth from Upware I cannot separate out any that are likely to belong to the Lower Greensand age. A number of vertebræ and fragments of paddles have occurred at Upware, amongst which Mr Seeley has recorded (*Cat. of Reptilia*, p. 76) VI. $\frac{6}{5}$ "a peculiar new species," and v. d. $\frac{6}{5}$, a dorsal vertebra, which is an instructive example, since it shews the epiphysis of the centrum fallen away, just as occurs in the backbones of living whales.

Also fragments of the paddle-bones have occurred at Upware which are remarkable for the shaft and epiphyses being often separated. The latter having fallen out now look like gigantic teeth or dermal spines, leaving the shaft as a double hollow cone like a dice-box, or rather like an hour-glass, for the two cavities are continuous. All these are probably 'derived' fossils belonging properly to the Upper Jurassic clays.

ICHTHYOSAURUS.

Some conical teeth with square fangs found at Upware and Brickhill are referable to the genus Ichthyosaurus. A fine specimen measures 60 mm. in length, of which the conical enamelled biting portion occupies only 15 mm.

CROCODILIANS.

Goniopholis? The crocodilian remains from Upware and Brickhill consist of teeth and a fragment of skull. The teeth are of the common crocodilian type, conical, terete, and slightly curved, with a pair of strong lateral crests, similar to the teeth of the *Goniopholis* of the Wealden.

A fragment of skull-bone exhibits the characteristic rain-pit surface-sculpturing of the crocodiles.

The lance-shaped teeth of *Dakosaurus* are of Kimmeridgian age, occurring at Upware only as 'derived' fossils.

PISCES.

NATIVE FISHES.

The remains of Fishes consist of teeth, fragments of jaws, and dermal spines, with occasional fragments of bones. For the most part they are rare fossils, but the hemispherical teeth of *Sphærodus* (known as *buttons* by the workmen) are amongst the most common and conspicuous fossils of the deposit. These are more or less mineralized with oxide of iron and phosphate of lime, principally the former, and the roots of the teeth have not served as centres for the formation of phosphatic nodules, such as is commonly the case with the *derived* teeth in the coprolite beds of the Suffolk Crag.

The close correspondence of the fish fauna in the Neocomian rocks of Upware, and St Croix in Switzerland, is very remarkable. Our *Sphærodus* teeth are identical with their *S. Neocomiensis*, Ag., our *Pycnodus* with their *P. Couloni*, Ag., and the jaws and teeth of *Gyrodus* are also similar in the two places. Again the Upware *Strophodus* corresponds well with that figured by MM. Pictet and Campiche (*Terr. Crét. St Croix*, Pl. XII. pp. 1—7).

The same *Sphærodus* occurs at Shanklin in the Isle of Wight, at Landeron, Neuchatel, Alais, and Auxerre. Now in these latter places derived fossils are, so far as I can learn, either unknown, or of very rare occurrence, and there seems to be no reason to doubt that the fishes' teeth likewise belonged to animals who lived in the Neocomian period (see *ante*, p. 17). Such must also be the case with many of our Upware specimens. On the other hand these palatal teeth shew no specific differences from those of *S. gigas*, Ag. from the Kimmeridge clay, nor do I think they could be distinguished and separated out if they were found mingled together. The fact of the undoubted occurrence of many Jurassic species in the Upware deposit affords good ground for the belief held by MM. Walker, Teall and others, that these are truly 'derived' fossils[1].

Under these circumstances I record both species, native and derived, from Upware, the proper Neocomian ones being, in my opinion, especially well witnessed to by the number of the specimens, and the scantiness of phosphate of lime either in their present mineral substance or in the form of concretions around the roots of the teeth.

SPHÆRODUS NEOCOMIENSIS, Agassiz.

(Plate I. fig. 4 *a—c.*)

Sphærodus Neocomiensis, Agassiz, 1843. *Poissons Fossiles*, II. 2, p. 216.

„ „ Pictet and Campiche, *Palæontologie Suisse*, St Croix, Vol. I. p. 72, pl. 9, f. 1—6.

We have already pointed out the resemblance of the teeth of this species to those of *P. gigas* of the Kimmeridge clay. MM. Pictet and Campiche write: "Je suis même très-embarrassé pour y trouver des différences appréciables," but add that perhaps the cap formed by the dentine and enamel is deeper and narrower in *S. gigas*.

These teeth are always isolated, and at Upware and Brickhill only the palatal type is known; but at Potton the sharper cutting incisor teeth are also found. Fragments of the jaw-bone occasionally occur still adhering to the tooth, but usually dentine and enamel only remain when the under side is cupped, and in some specimens the dentine also has disappeared, leaving only the saucer-like enamel cap, shaped like an acorn cup. These palatal teeth were known as 'buttons' amongst the workmen, for whom they were

---

[1] As mentioned before, I have no knowledge of any specimen of *Sphærodus gigas* or *Strophodus reticulatus* having been found in the Jurassic rocks of this neighbourhood (Cambs. and Bucks.).

K. 6

actually manufactured up into coat buttons in considerable numbers (!).

Localities. Upware, Brickhill, Potton, Shanklin (I. W.), Farringdon (Berks.).

S. Europe. Dept. de l'Aube, Alais, Auxerre, Landeron, Neuchatel, St Croix.

STROPHODUS, sp.

Strophodus reticulatus, Agassiz?

Psammodus reticulatus, Walker, *Geol. Mag.* Vol. IV. p. 310 (see *Psammodus reticulatus,* Ag. *Poissons Foss.*Vol.III.p. 123, pl.17,I. From the clay at Shotover (Kimmeridge clay).

This species is undistinguishable from the *Strophodus reticulatus,* Ag. of Shotover, and also from the *S. subreticulatus,* Ag. of Soleure, according to MM. Pictet and Campiche. Still there can be little doubt that when better specimens are found to shew the arrangement of the teeth in the jaws we shall find evidence of the specific distinction of the Neocomian and Kimmeridgian species, and the former might then very appropriately celebrate the great Palæontologist who first described it by bearing the name *Strophodus Picteti.*

Localities. Upware, Brickhill, Potton.

S. Europe. St Croix, Switzerland.

PYCNODUS COULONI, Agassiz.

(Plate I. fig. 5, *a, b.*)

Pycnodus Couloni, Ag. 1843. *Poissons Fossiles,* II. pt. 2, p. 200.

„   „   Pictet and Campiche, *Terr. Crét.* St Croix, p. 57, pl. 7, fig. 5—17.

„   „   de Loriol et Gillieron, *Pal. Landeron,* pl. 1, 5—7.

Pycnodus gigas, Walker, *Geol. Mag.* Vol. IV. p. 310 [and *Pal. Soc.* in Lycett].

With this species there exist difficulties just similar to those we meet with in the Sphærodus teeth. Agassiz writes: " Elles sont tellement semblables au *P. gigas* du Portlandien qu'il est presque impossible de les distinguer."

Localities. Upware, Brickhill, Potton.

S. Europe. St Croix, Landeron, Neuchatel, Thoiry, Alle-moque (dept. de l'Ain). Urgonien and Neo-comian (de Loriol and Gillieron).

OTODUS, sp.

A short and stout hastate tooth. The root is massive but not expanded; two slight enamelled eminences on its sides represent the lateral denticles of the genus.

Length. 1 inch.

Some deeply biconcave sharks' vertebræ, measuring 1—3 inches in diameter, are likely to belong to the same species.

Localities. Upware, Brickhill, Potton.

Besides these well-marked types, several other species of fishes occur at Upware and Brickhill, and have been enumerated amongst the 'derived' fossils, though some of them may be proper to the bed. There is a particularly fine dorsal spine of *Asteracanthus ornatissimus* from Upware in the Woodwardian Museum, measuring 7 inches in length; and the jaws of *Ischyodus Townshendi* are found in fine condition both at Upware and Potton. Other species of Chimeroid jaws also occur.

II. INVERTEBRATA.

CEPHALOPODA.

(See *ante*, p. 18.)

Belemnites (*Actinocamax*) *pistilliformis*, Blainville, *Mémoir sur les Belemnites*, p. 98, pl. v. figs. 14 and 15, 1827. [Not *B. pistilliformis*, Sby. *Min. Con.* 1828.]

    ,,   *jaculum*, Phillips, *Geol. Yorkshire*, I. pl. iii. fig. 1.

    ,,   *pistilliformis*, Duval, *Journ. Belem. des Terr. Crét. infr. de Castellano*, pl. 8, figs. 10—16.

    ,,   *Ewaldi*, Strombeck, *Deutsches Geolog. Gesell.* 1861, p. 34.

    ,,   *pistilliformis*, Ooster, *Pet. des Alpes*, I. Cephalopoda, pl. 2, figs. 9—11, p. 21.

    ,,   *pistilliformis*, Renevier, *Terr. Neocomian des Voirons*, pl. 1, figs. 1—4, p. 5.

[*Belemnites subfusiformis, auctorum* is a variety of this species. See *infra.*]

THIS species belongs to a group of Belemnites characterised by the soft and powdery nature of the alveolar region of the guard, so that commonly this part is not preserved. Further, in consequence of the outer layer being more friable than the inner, a delamination occurs in successive coats—'en lames en retrait.' A perfectly fusiform structure is thus produced, being a *guard*, wanting its alveolar portion and its phragmocone. The Genus *Actinocamax* (Miller and Voltz) was erroneously grounded upon specimens in this condition, and the name has consequently been rejected. Dr Schlüter has, however, lately revived the name, and I think it may be retained with advantage to denote a well-marked group of Cretaceous Belemnites, including *Belemnites pistilliformis, subfusiformis, minimus, attenuatus, ultimus,* and *semicunaliculatus*[1].   All these species are very nearly allied to one another, and specimens are often difficult to determine.   Another characteristic feature of the group *Actinocamax*, is the presence of a double pair of lateral grooves running along the guard, but this is not so constant nor so characteristic a feature as the one first mentioned.

This species is well known from Speeton in Yorkshire, under the name *Belemnites jaculum*, Phill.   Elsewhere in Europe the name *jaculum* is scarcely known, Phillips's figure being too rough for identification (Goldfuss).   I cannot find any character to distinguish it from *B. pistilliformis*, Blainville.

Herr von Strombeck tells me he considers his species *Ewaldi* to be the same as *jaculum*, Phillips.

Measurements, etc.   In the Woodwardian Museum are three specimens, all of them of rather slender, elongated type, regularly but not strongly fusiform.

Length, 2¾ inches; breadth at centre, 8½ mm.; breadth at ½ in. from proximal end (*Actinocamax*), 7 mm.; breadth at ½ in. from apex, 7 mm.

Localities.   Upware, Speeton, Tealby (Mid. Neocomian).

N. Europe.   Elligserbrink (Hilsthon); Kirchdorf, Oberer Hils (Gottingen Museum).

S. Europe.   Varappe, Alpine Cretaceous (Ooster), Mont Salève (Mid. Neocomian), Voirons, St Croix, Castellana.

BELEMNITES PISTILLIFORMIS, var. SUBFUSIFORMIS, d'Orbigny.

(Plate I., fig. 6, *a—d*).

Belemnites pistilliformis, var. *subfusiformis*, Raspail, 1829, *Hist.*
       Nat. des Belemnites, pl. VIII, fig. 93, p. 55 ?
   „   *subfusiformis*, d'Orbigny, *Terrains Crétaces*, I. p. 50,
       pl. 4, fig. 9, also *Supplement*, p. 24.
   „   *minimus*, de Blainville, 1827.   *Mém. sur les Belem-*
       nites, t. v. f. 5, 6; t. IV. f. 1 (non Sowerby).

---

[1] *Belemnites hastatus* of the Oxford clay, and *B. clavatus* of the Lias, occur in a similar condition, but the double pair of lateral grooves are not well developed in these species.

Belemnites subfusiformis, Duval Jouve, *Belem. Terr. Cret. Infr. Castellana*, pl. IX.

    „   *subfusiformis*, de Loriol, *Descr. des fossiles, Neocom. moyen du Mont Salève*, pl. I. fig. 1.

    „   *semicanaliculatus*, Pictet and Renevier, 1858, *Palæont. Suisse, Foss. du Terr. Aptien.* p. 19, pl. III. fig. 1.

There is some doubt as to the real type of the Belemnite thus named by Raspail, Quenstedt (*Petrifactenkunde Deutschland*, Cephalopoda), believing it to be a Jurassic (Weiss Jura) species, different from that figured by d'Orbigny. We therefore refer to the figures by Duval Jouve and d'Orbigny, as illustrating the current idea of the species, being unable to settle the question raised by Quenstedt.

This form passes by easy gradations into *Belemnites pistilliformis* (Blainville), as was recognised by d'Orbigny in his supplement (*Terr. Crétaces*), and the two so-called species have been united under one name by Pictet and Renevier. In the North German collections I find them still kept distinct, and I think the name *subfusiformis* may with advantage be retained as a varietal name for those forms with a decided ventral furrow at the alveolar end. Further, this type is as a rule less swollen but more regularly fusiform than the true *pistilliformis*, and the lateral furrows are less distinct.

Measurements. Length, 2⅜ inches; greatest breadth (just below the middle), 10 mm.; breadth at lower end of proximal furrow, 8 mm.; breadth at ¼ inch from apex, 7½ mm.

Localities. Upware, also at Hythe (Woodwardian Museum).

N. Europe. Elligserbrink, Hilsthon (Brunswick Museum).

S. Europe. · Alp. Mar. Claro (Neocomian), Castellane (d'Orb.),   Perte du Rhone, Aptien supr. (Pictet and   Roux).

BELEMNITES SUBQUADRATUS, Roemer.

(Plate I., fig. 8, *a, b*.)

Belemnites subquadratus, Roemer, *Norddeutsch Oolitengebirge*, 1836, p. 166, t. 16, f. 6.

    „   „   Ooster, *Fossiles des Alpes*, Ceph. p. 24.

Belemnites subquadratus, d'Orbigny, *Terr. Crét.* Sup. p. 12, t. 6, f. 1—4.

„   „   Quenst. *Pet. Deutschl.* 1849, Vol. I. p. 462, t. 30, f. 26, 27.

„   *lateralis. Cat. Foss., Museum of Practical Geology*, 1878, p. 23 (Tealby series).

One would hardly have been able to identify our small Brickhill Belemnite as the one figured by Roemer and d'Orbigny without the help of other specimens. Ours is a small, simply-tapering fragment of a guard, worn and flattened on its ventral side, and with a very excentric axis of growth as its most marked character. It is identical with some small Belemnites common in the Hils conglomerate of Goslar, Salzgitter and Schoeppenstadt (Brunswick), where I have collected examples, and in these localities it occurs associated with larger and more typical forms.

Measurements. Length, 3½ cm.; breadth, side to side, 7 mm.; breadth, back to front, slightly less (6¾ mm.).

Localities. Brickhill (W. M. unique), Farringdon (W. M.), not uncommon.

N. Europe. Hils conglomerate of Schandelahe, Bredenbeck, Achim.

S. Europe. Bernese Alps, Wassy, Haute Marne, Neocomian (d'Orb.). Römer's type is from the Jurassic rocks.

BELEMNITES UPWARENSIS, n. sp.

(Plate I., fig. 7, *a—d.*)

A most remarkable little Belemnite utterly dissimilar to any other Belemnite that I know in several of its characters, namely in the markings over its whole surface, the structure disclosed by transverse section, and the characters of the alveolar end of the guard. So that although only a single specimen appears to have been found, I have no hesitation in recording it as a new form, under the name *Belemnites Upwarensis.*

Description. Guard short and stout, conoidal[1], being a little de-

---

[1] For Terminology, see Phillips, *British Belemnites*, Pal. Soc. page 30.

pressed by a dorsal flattening; it expands slightly at the proximal end : distal end submucronate, mucro short and stout.　No furrows. The *Surface* is completely covered with ridges and striæ, some of the former corresponding in position with the structural radii seen in transverse section : the striæ are smaller, and much more numerous (about 10 striæ to 1 mm.).

The proximal end shews a shallow cavity much like the calyx of a coral, the constituent radii standing out clearly in pencil-groups suggestive of coral septa.　The expanding of the guard at this region shews that this proximal cavity was at, or near to, the alveolar cavity, but it obviously is not simply that cavity itself but is the result of destroying agents acting upon original peculiarities of structure.　Whereas in *Actinocamax* we have a group of Belemnites with the outer layers more and more friable in the neighbourhood of the phragmocone, here the contrary seems to be the case, so that a hollowing out of the upper end of the Belemnite has occurred.

A transverse section shews the contour to be nearly circular, rendered somewhat irregular by the slight dorsal flattening.　The characteristic exogenous structure is well seen, with few and strong growth circles and unusually well-marked radii.　These radii converge to the centre, most of them being continuous so that the rays become numerous and closely packed towards the centre, but as the more central part is approached they become irregular, curving, and wavy, converging towards and uniting with one another in a manner recalling the section of a Palæozoic *cyathophyllum*.　In ordinary Belemnites the radii run straight to the centre—(Compare Fig. in Phillips's *British Belemnites*, Pal. Soc., p. 14).

The axis of the guard is subcentral.　Phragmocone unknown.

Measurements.　Length, 23 mm.; breadth near middle, 7½ mm.; lateral diam. at proximal end, 8 mm.; dorso-ventral diam. at proximal end, 8 mm.; dorso-ventral diam. at middle, 7½ mm.　The axial line meets the diameter excentrically, so as to mark off radii of 3 mm. and 4 mm. respectively.

AFFINITIES AND DIFFERENCES.　*Belemnites Upwarensis* is well separated from all other species by the ornamented character of its whole surface.　I believe these surface striæ to be the true surface structures produced by the formative membrane and not due to decomposition as in *B. acuarius macer*, Quenst., for the striæ are very regular and are best developed where the surface has suffered least in fossilization.

The form too is sufficient to distinguish it from other Cretaceous Belemnites, while yet other specialities are probably indicated by the condition of the curious proximal cavity. We find therefore in this Belemnite several such important special features as suggest that we have here the type even of a new genus of Belemnitidæ.

Locality. Upware.

AMMONITES CORNUELIANUS, d'Orbigny.

(Plate I., fig. 9, *a—c.*)

Ammonites Cornuelianus, d'Orb., *Terrains Crétaces*, I., pl. 112, f. 1, 2, p. 364.

„ „ Pictet and Roux, *Mollusc des Gres verts*, pl. V., f. 4, *a, b,* p. 55.

„ „ Ooster, *Fossiles des Alpes*, IV., Cephalopoda, p. 132 (Bernese Alps).

A variable Ammonite, belonging to a variable group—the *Nodosocostati* (P. and C.). It is closely allied to our Atherfield *Am. Martini* (d'Orb.) with which it has been bound up by MM. Pictet and Campiche in the above-named group. Quenstedt unites this species and *Martini* with *verrucosus* d'Orb., (Quenst. *Petrifactenkunde*, Vol. I., p. 136). *Am. Martini* is a thicker form, less open, than our species, and the spines are better developed in the later whorls. In the Woodwardian specimen the spines are well developed on the younger whorls but become obtuse and rather obscure further on.

Two varieties occur at Upware, a rotund variety (A) (round whorled) and a flatter whorled form (B) ; the latter approaches *A. Milletianus,* d'Orb.

The sharply defined, obtuse, square-sectioned ribs, especially over the back, are very characteristic of this group of Ammonites.

Our three best specimens (Woodwardian Museum) measure about 3 inches, 1½ and ¾ inches respectively.

Localities. Upware.

N. Europe. Louvenmont (Göttingen Mus.).

S. Europe. Perte du Rhone (Aptien supr.), Switzerland, Bernese Alps, Basses Alps.

"A characteristic Aptien species," (Pictet and Renevier, *Perte du Rhone,* p. 21).

AMMONITES DESHAYESII, Leymerie.

(Plate III., fig. 1.)

Ammonites Deshayesii, Leymerie, 1842, *Mém. Soc. Geol. France.* t. v., pl. 17, f. 17.

„ *consobrinus,* d'Orbigny, *Terr. Crét.,* t. 47.

„ *fissicostatus,* Phillips, *Geol. Yorkshire,* t. 2, f. 49.

In the German collections this form is usually included in *Ammonites noricus* of Schlotheim (the original types of which came from the Elligserbrink Neocomian clays). But taking Roemer's description and figure of that species as our guide (the original diagnosis being too imperfect to serve), it will be generally admitted that the two names, *Deshayesii* (Leym.) and *noricus* (Schl. ? and Roemer) should be retained for their respective types. They are however very nearly allied, and a connecting chain of transition specimens could, I am convinced, without difficulty be arranged. *Ammonites Deshayesii* has fewer ribs and coarser than *A. noricus,* and no tubercles are developed along the outer border where the ribs are deflected to pass over the back. Again the ribs in *Am. Deshayesii* pass regularly over the back, simply flattening out and deflecting forwards, leaving no such pseudo-keel as we find in *A. noricus.*

Only a single fragment is known from Upware, belonging to a variety with broad obtuse ribs.

It is an interesting fact that the more typical coarser-ribbed form of *Ammonites Deshayesii,* Leymerie, occurs at Upware as a "derived" fossil!

Localities. Upware (Woodw. Mus.), Speeton (Upper Neocomian), Atherfield.

N. Europe. Brunswick.

S. Europe. Paris Basin, l'Aube.

Ammonites, sp. ?.

(Plate I., fig. 8.)

A fragment in the collection of Mr J. F. Walker, M.A., of
Sidney Sussex College. It was a simple round-backed Ammonite
with very numerous ribs (30 to 1 inch). Diam. of whorl ⅜th inch.
I have not been able to identify this species, and it is too imperfect
for further working.

Locality. Upware (Coll. Mr J. F. Walker, M.A., F.G.S.).

ANCYLOCERAS HILLSII, Sowerby, sp.

(Plate II.)

Scaphites Hillsii, Sowerby, *Geol. Trans.* IV., p. 339, pl. XV., f. 1, 2.

Only one specimen of this species is known from Upware
(Woodw. Mus.). It is a fine shell just so far advanced as to have
completed its *Crioceroid* stage of growth. The aperture here is
rounded and transversely oval, whereas the section of the inner
whorls and of the mature *Ancyloceras* part of the shell is nearly
square. The ribs become very faint over the last quarter-whorl of
our specimen.

The species differs from *A. gigas* Sowerby in its whorls being
more rounded, and enlarging less rapidly; also in its more simple
rounded ribs in the Crioceroid stages. From *Crioceras Asterianus*
d'Orbigny (*Terr. Crét.* I., pl. 115, bis. f. 3, 4, 5) our shell differs in
its more distinct ribbing, by its whorls enlarging less rapidly, and
in the smaller dorso-ventral diameter of its aperture.

Localities. Upware, Hythe, Lympne, Maidstone, Godalming(?).

GASTEROPODA.

(See *ante*, p. 19.)

APORRHAIS (TESSAROLAX) GARDNERI.

(Plate III., fig. 2, 2 *a*.)

Aporrhais (*Tessarolax*) *Gardneri*, sp. nov.
Tessarolax, n. sp. J. S. Gardner, ' Cretaceous Gasteropoda,' *Geol.
Mag.*, 1880, Vol. VII., pl. iii., f. 2, p. 50.

" This fragment possesses all the characters of a *Tessarolax*,
and is similar in shape to *T. retusa*, the small extent of the region
in front of the keels giving the shell a truncated instead of a pear
shape.   The species is evidently finely striated, and the keels
appear to have been slightly tuberculated as in *T. Fittoni*.   It
was a larger shell than either of the Gault or Neocomian forms ;
but less than the grey chalk *T. oligochila*.   It is scarcely sufficiently perfect to furnish specific characters."
Locality.   Upware (Coll. Mr J. F. Walker, unique).

APORRHAIS (TRIDACTYLUS) WALKERI, Gardner.

(Plate III., fig. 3.)

Tridactylus Walkeri. J. S. Gardner, *Geol. Mag.*, Vol. VII., pl.
iii., f. 4, p. 50.

" Shell elongate and pupæform ; whorls irregular, inflated, not
forming a regular cone ; possessing a strong central keel, a second
and less prominent keel in front, and two partly concealed sutural
keels.   The body-whorl has two nearly equal keels (slightly
diverging towards the outer lip, the front one being the less prominent), and three smaller keels which are arranged above, below,
and between the primary keels.   All the keels seem to have been
tuberculated, and the spire was probably ribbed, as in *T. Grif-
fithsii*, which it generally resembles.   Length of fragment, 22 mm. ;
diameter of body-whorl, 12 mm.

"This unique fragment possesses great interest, since the only two cretaceous species previously known were from the Gault. It is remarkable for the irregularity of its growth, the third whorl bulging to the left and the second to the right; this tendency perhaps preceded the more regularly pupæform shapes of the Gault species."

Locality. Upware (Coll. Mr J. F. Walker).

SCALARIA KEEPINGI, Gardner.

Scalaria Keepingi, Gardner. *Geol. Mag.*, VII., 1880, p. 54, pl. iii., fig. 7.

"Whorls inflated, about twice as wide as high; ribs of each whorl coarse, rounded, very irregular, flexuous; striæ very fine and scarcely perceptible under a glass; suture not visible.

"This species, while resembling Neocomian forms in other respects, differs from them all both in the nearly complete absence of striæ, and also by its concealed sutures."

Length, 19 mm.

Locality. Upware (Coll. Mr J. F. Walker, M.A.; unique).

CERITHIUM NEOCOMIENSE (d'Orb.), Forbes.

(Plate III., fig. 5, *a, b.*)

Cerithium Neocomiense, d'Orb.? *Terr. Crét.* II. pl. 232, f. 8—9.
  „   „   Forbes. *Quart. Jour. Geol. Soc.*, Vol. I., 1845, p. 351, pl. IV., f. 8.

A single prominent keel upon the upper whorls; two keels exposed upon the body-whorl.

Localities. Upware (Coll. Mr J. F. Walker), Atherfield.

CERITHIUM MAROLLINUM, d'Orbigny.

(Plate III., fig. 6.)

Cerithium Marollinum, d'Orb., *Terr. Crét.*, p. 353, pl. 227, fig. 2, 3.

Localities. Upware, Tealby, E. Shalford (Surrey).
  S. Europe. Marolles (Aube), Yonne (Neocomian).

94 INVERTEBRATA.

NERINÆA, sp.
(Plate III., fig. 7.)

A tall turreted shell resembling the *Turitella nerinœa* of Geinitz. We know of only two specimens from Upware, and perhaps another from Atherfield. These indicate a variable species, and do not afford sufficient materials for specific identification or diagnosis. The whorls are concave, being strongly margined at each suture. One specimen shews some spiral striations, but the other is smooth.

Localities. Upware, Atherfield (?).

NERINÆA TUMIDA (n. sp.).
(Plate III., fig. 8.)

Shell thick, elongated, tumid, many-whorled, smooth. Whorls narrow, simple, convex; sutures rather deep, very transverse. Base much contracted, umbilicus wanting. Mouth sub-quadrate; siphonal canal rather prominent, slightly tortuous; columellar lip straight. Two broad faint grooves seen in places upon the inside cast indicate slight spiral thickenings of the outer wall.

Measurements of type (see figure): Length (of fragment) 3¼ inches; breadth at top, 22 mm.; at base, 35 mm.; height of penultimate whorl, 13 mm.; angle of spire, 13°.

Only two specimens are known, but these indicate a well-marked species, the general tumid, almost pupoid appearance of the shell being very distinctive. The convexity of the whorls and contracted base distinguish it from *N. Banga*, d'Orb., and from *N. depressa*, Voltz., to which it has most general resemblance. It is separated by the absence of an umbilicus.

Locality. Upware.

TROCHUS, n. sp.
(Plate III., fig. 9, *a*, *b*.)

A trochiform shell allied to *T. Oosteri* (P. and C.) of the Aptien Inferieur of St Croix. It is regularly pyramidal, or with the apex slightly produced; base nearly flat, gently convex; no umbilicus.

Whorls 7 or 8, each with a prominent keel forming its anterior margin; sutures deep and angular. Between the keel and the suture the whorls are ornamented with two or four tuberculated secondary ridges, the two median ones being strongest and most constant.

Base ornamented with spiral striations, of which the strongest are next the border-keel.

Mouth triangular, slightly effuse below.

Only a single specimen is known, in the cabinet of Mr J. F. Walker, M.A., and this is not sufficiently well preserved to serve for perfect specific diagnosis.

Measurements. Height, 12 mm.; breadth, 13 mm,; apicial angle, 80°.

Locality. Upware (Collection Mr J. F. Walker, M.A.).

LITTORINA, sp.

Nearly allied to *L. Upwarensis*, but much shorter, being nearly as broad as long.

Locality. Upware (Woodw. Mus., unique).

LITTORINA UPWARENSIS, sp. nov.

(Plate III., fig. 10, *a, b.*)

Shell thick, pyramidal, with oblique base, longer than broad, ornamented with tuberculated spiral ridges. Sutures deep; whorls 6 or 7, convex, ornamented with two tuberculated keels, of which the more posterior is the stronger and more coarsely nodulated. This keel occupies the centre line of each whorl in the spire. Posterior to this the whorls are usually slightly concave, and plain, but an additional ridge, with tubercles, may be developed along the sutural border.

The body-whorl has numerous (9 or 10) spiral ridges upon its sides and base, the posterior two or three being distinctly tuberculated, the remainder less so.

The base is convex and oblique.

Mouth rhomboidal; the columellar lip slightly produced.

Besides the spiral ornament already described, the whole shell is covered with a delicate cross striation, running in parallel curved lines between the ribs. The shell varies considerably in the development of its spiral ridges and tubercles, the median range of tubercles being sometimes prominently developed.

Measurements. Length, 23 mm.; breadth, 17 mm.; apicial angle, 55°.

Locality. Upware.

LITTORINA CANTABRIGIENSIS, sp. nov.

(Plate III., fig. 11, *a, b.*)

Shell pyramidal, higher than broad, with oblique base; ornamented with numerous tuberculated spiral ridges, and cross striations. Spire regularly conic, consisting of six or seven whorls separated by narrow but deep sutures. Body-whorl with a distinct shoulder, in front of which the shell is rather scanty. A tuberculated keel runs along the shoulder, and a second keel, almost as strong, is found next in advance; the next ridge is much less developed, and then comes the sutural ridge, or a row of tubercles.

Anterior to the keels are four or five spiral ridges, where the tubercles are but faintly developed. In the spire the whorls are flat, ornamented with four tuberculated ridges, one of which may disappear.

Base scanty, with a small umbilical fossa. Mouth rhomboid.

Measurements. Height. 18 mm.; breadth, 14 mm.; apicial angle, 60°.

The ornament structures are well developed in this shell, the cross striæ between the ridges being coarser and more irregular than in the other Upware species.

Locality. Upware (very rare). Woodw. Mus.

N. Europe. Ocker, near Goslar (Hils Conglomerate).

LITTORINA (TURBO) VARICOSA, sp. nov.

(Plate III., fig. 12, *a, b.*)

A turbinate shell, longer than broad, ornamented with spiral ribbing and tubercles, interrupted at intervals by prominent varices. The base is short and rounded; imperforate. Spire composed of five or six whorls separated by rather broad and deep sutures.

The body-whorl is convex and rounded with spiral ornament belonging to the usual two sets—(a) the spire-system, consisting of three rows of tubercles which are more or less connected so as to form tuberculated ridges; and (b) the anterior or basal set of seven or eight spiral ridges which are faintly tuberculated. The whole shell is covered with delicate cross striæ.

Mouth subquadrate.

The varices are simple, convex and very prominent; four are seen at irregular periods upon our type specimen.

Measurements. Length, 16 mm.; breadth, 12 mm.; apicial angle 55°.

Locality. Upware (rare).

TURBO REEDI, n. sp.
(Plate III., fig. 13, a, b.)

A small, stout, turbinate shell; round whorled, but strongly keeled; striated and reticulated. Sutures deep, whorls five or six, convex, with two (or three) strong keels upon the anterior part; posterior half of each whorl concave.

Body-whorl with numerous (about fifteen) spiral ribs upon its sides and base.

Base convex, with a small umbilicus.

The whole surface is ornamented with a delicate spiral striation and oblique cross striations producing a beautiful decussated or granulated appearance.

Measurements. Height, 13 mm.; breadth, 11 mm.; apicial angle, 70°.

The shells vary in depth of suture and in the number of spiral ribs exposed over the spire.

Locality. Upware.

PATELLA, sp.
(Plate III., fig. 14.)

Too imperfect for identification or description. Internal view only known. See figure.

K. 7

PLEUROTOMARIA GIGANTEA, Sowerby?

Pleurotomaria gigantea, Sby., *Trans. Geol. Soc.*, 2 Ser., tom. IV.,
pl. XIV., f. 16.

"   "   Pictet and Campiche, *Pal. Suisse, Terr.*
Crét., St Croix, Vol. III. p. 433.

There are two types of this large pyramidal *Pleurotomaria*,
namely, the more elevated, flat-whorled type with numerous volu-
tions found at Atherfield, and a more depressed broader-whorled
type common at Hythe ; in the latter also the sides of the whorls
are more arched, and expand at their base. Again, in the Hythe
form the spiral ornament lines are coarser and more prominent,
and the line of the apertural slit is situated at or below the
centre of each whorl. In the Atherfield type this line is usually
slightly above the medial line.

Our Upware shell is nearest to the Hythe type, but as only
the inside cast is known the identification is not certain.

Localities. Upware, Hythe, Atherfield (?).

S. Europe. St Croix.

PLEUROTOMARIA RENEVIERI, n. sp.

(Plate IV., fig. 1, *a, b.*)

Pleurotomaria Anstedi, *Cat. Fossils Jermyn Street Museum*,
1878, p. 21.

[not   "   "   E. Forbes, 1845. *Q. J. Geol. Soc.* I.
p. 349, pl. 5 (13), f. 1.]

["   "   "   of Pictet and Campiche, *Pal. Suisse*,
p. 435.]

In his description of *Pleurotomaria Anstedi*, Prof. E. Forbes
writes: "spire somewhat depressed and very obtuse," and his
figure corresponds well with the description; also the margin of
the shell is rounded. The Upware shells and that figured by
MM. Pictet and Campiche are pointed and their edges sharp.

These characters are very constant in the genus *Pleurotomaria*, and our specimens shew no variation in the directions
of Prof. Forbes' type.

Description. Shell thick, regularly conical, broader than high.
Base convex, apex acute, whorls few (six or seven), with convex

sides, ornamented with a strong thickening at their anterior margin and two ridges bordering the sinus-band. A cancellate ornament is developed upon the marginal thickening and above the sinus-band by the intersection of well-marked spiral striations and cross striæ. Elsewhere the sides of the cone are ornamented only with transverse striæ. Base cancellated with spiral ridgings and cross striæ. Umbilicus moderate. Mouth rhomboid.

Specimens vary in the development of the striæ and costæ.

From *P. elegans* (d'Orb.), and *P. Cassiana* (d'Orb.), this species is distinguished by the absence of spiral ridges upon the whorls in front of the slit. The general proportions of the shell are, as kindly pointed out to me by Prof. Renevier, similar to those of the internal casts from Mont Salève, described by M. de Loriol as *P. Saleviana*.

Localities. Upware.

N. Europe. Berklingen, Brunswick.

S. Europe. St Croix.

Measurements. Height, 20 mm.; breadth, 24 mm.; apicial angle, 85°.

PLEUROTOMARIA FERRUGINEA, sp. nov.

(Plate IV., fig. 2, *a*, *b*.)

Shell thick, pyramidal, elevated, slightly higher than broad, consisting of nine or ten whorls which are angular so as to give the shell a stepped appearance. Sinus-band mesial in each whorl, occupying the axis of the keel. Base moderately convex, nearly smooth, but with faint spiral ornament striations, and distinct lines of growth. Mouth trapezoidal, somewhat effuse below; margin thickened. Umbilicus small, slit-like.

The shell is ornamented at the sides with spiral striations crossed by delicate striæ, and by lines of growth; about thirty of the spiral ridges are present upon the later whorls.

Measurements of type. Height, 3¼ inches; breadth of base, 3 inches; breadth of sinus-band, 1 mm.; angle of spire, about 55°.

This species belongs to the *Gigantea* group, but differs from *gigantea* in its more convex and nearly smooth base, also in the

7—2

mesial angle of the whorl. From *P. Delahayesii*, d'Orb., it is
distinguished by its taller form, small umbilicus, distinct spiral
striations, and mesial sinus-band. *P. Favrina*, de Loriol, has
a characteristic depression around the umbilicus.

Locality. Potton.

LAMELLIBRANCHIATA.

(See *ante*, p. 19.)

OSTREA (EXOGYRA) COULONI (Defrance), d'Orbigny.

Gryphea Couloni, Defrance, 1821. *Dict. Sci. Nat.* XIX. p. 534.
E. subsinuata, Leym. 1842. *Mém. Soc. Geol. de France*, tom. V.
          pp. 11, 16, 17, Pl. 12, figs. 3—7.
Ostrea Couloni, 1846. d'Orb., *Terr. Crét.* pl. 467.
    ,,   ,,   1858. Pictet and Renevier, (*pars*), *Pal. Suisse*,
          Terr. Aptien, p. 138.
    ,,   ,,   1861. de Loriol, *Mont Salève*, p. 110.

MM. Pictet and Renevier have united this species with *O.*
(*Exogyra*) *sinuata* (Sow.), (*Foss. du Terr. Aptien, Pal. Suisse*,
p. 138), and in N. Germany the two names are usually regarded
as synonymous. But de Loriol (*Foss. du Mont Salève*) doubts
the wisdom of that arrangement, and I agree with him that the
two forms may well be kept separate though they are closely
allied. I find that where *Exogyra sinuata* (Sow.) occurs there
E. Couloni is also found — but the converse does not always
obtain—as at Upware, Farringdon and Mont Salève.

The true *sinuata* (Sby.), may be called *Ex. Couloni var. sin-
nata* (Sby.).

Measurement. Length of largest specimen, 2½ inches.

Localities. Upware, Brickhill, Speeton (Lower Neocomian),
        Tealby, Potton.

N. Europe. Brunswick, Hils Thon, and Hils Conglomerat.
S. Europe. Perte du Rhone, Mont Salève, &c.

EXOGYRA CONICA, Sby., *vars.* including (Plate IV. fig. 3,
a, b, c)

Chama conica, Sby., 1812. *Min. Con.* pl. 26, fig. 3.
Exogyra conica, Sby., *Min. Con.* t. 605, figs. 1—3.
„ *undata*, Sby., *Min. Con.* t. 605, figs. 3, 4.
„ *spiralis*, Roemer, 1839. *N. Ool.* p. 65 (*pars*).
„ *undata*, „ 1841. *Nordd. Kreid.* p. 47, No. 7.
Ostea Tombeckiana, d'Orb., *Terr. Crét.* p. 701, pl. 467, figs. 4—6.
„ *conica*, „ „ p. 726, pl. 478, figs. 5—8.

Our Upware forms are usually known as *Exogyra Tombeckiana* in England and France, but they are called *E. spiralis* in the German Museums. Prof. Morris and Mr Blake record *E. spiralis* as an oolitic species (Portland Oolite and Coral Rag), and Roemer in his original description gives both the Corallien and the Elligserbrink beds (Neocomian), for its horizons. The Jurassic and Cretaceous shells do indeed closely correspond, and they all belong to the same group. *Exogyra nana* is especially near to the variety *spiralis*, and others very nearly resemble even *E. virgula*, some of the specimens having been mistaken for this latter species and recorded as such in the published lists of Potton Shells.

Three well-marked types of this species occur at Brickhill, namely:

(1) *Exogyra conica*, Sby., *var. Tombeckiana*, d'Orb. This is the true *Ostrea Tombeckiana*, a small smooth species, with crenulated lip.

(2) *Exogyra conica*, Sby., *var. spiralis*, Roemer. A variety having its upper valve ornamented with regular and distinct imbrications; and

(3) a striated variety.

Localities. Upware, Potton, Brickhill, Farringdon, Blackdown.

N. Europe. Haverlah, Gevensleben; Schoeppenstedt, Hils Conglomerat.

S. Europe. Mont Salève, Perte du Rhone, etc.

OSTREA (GRYPHEA) DILATATA, (Sby.) ?

Ostrea dilatata, Sowerby, *Min. Con.* t. 149, fig. 1.
„   *controversa*, Roemer, *Nordd. Ool.* pl. 4, fig. 1.

I cannot detect any characters of distinction between the Upware and Potton Gryphites and the Oxfordian *G. dilatata*. Both the typical and the expanded forms occur, the latter most commonly. They measure respectively 5 × 3 inches, and 4 × 4 inches.

There being so many 'derived' fossils in the Upware deposit the suggestion is spontaneous and natural that this shell is of that nature having been washed out from the Oxford Clay of the old shore line and buried in again with true Neocomian fossils. Such was the opinion of Mr J. F. Walker; but I am convinced with Professor Seeley that this is not the case, but that this oyster is really a native of the deposit. Both at Upware and Potton the condition of the shell is precisely that of the true natives and quite different from that of any of the derived shells.

Moreover, although the Oxfordian Ostreæ are almost invariably covered with *Episites* (*Serpula, Ostrea,* and *Polyzoa*), there are no such Jurassic Episites upon the Upware Ostreæ.

But although the Oxford Clay and Upware shells are thus inseparable, we cannot, in my opinion, be justified in coming to a certain decision that they are the same species until they are found in some of the intervening rocks of the gap which now separates them in geological time.

It is a curious point that, excluding the *Ostrea frons* type, the Upware Ostreidæ would lead the palæontologist to believe he was amongst the Jurassic rocks—Lower Kimmeridgian or Upper Oxfordian or rather Ampthill Clay. For besides the species now under discussion, there is in the Woodwardian Museum a shell identical in shape with *Ostrea deltoidea*; and the little *Exogyra conica* and *E. spiralis* are undistinguishable from *E. nana* of the Kimmeridge Clay and Coral Rag.

OSTREA FRONS (Park.), *var.* MACROPTERA (Sowerby).

Ostrea macroptera, Sowerby, *Min. Con.*, pl. 478, figs. 3—5.

Ostrea frons, Parkinson (*Org. Rem.* III., pl. 15, fig. 4), is the type of a beautiful group of oysters which first appeared in the Upper

Jurassic rocks, but which are very characteristic of the Cretaceous series. A number of names have been given to different varieties such as *O. diluviana, macroptera, carinata, larva* and *rectangularis*, all of which names distinguish well-marked types more or less localized in their geological and geographical distribution, but which are found to pass into one another through intermediate forms when a large series is examined. The relation of the types to one another is indeed well illustrated by the arrangement under Prof. Geinitz in the Dresden Museum where the well-marked types are arranged on tablets and a large tray of intermediate forms is also exhibited.

The Upware forms belong decidedly to the variety *macroptera* (Sby.), which is distinguished by its umbonal region being expanded to form a wing, especially on the posterior side.

Localities. Upware, Brickhill, Potton, Farringdon.

N. Europe. Hils formation, Schöppenstedt.

S. Europe. France and Switzerland.

Ostrea frons, Parkinson, var. *carinata*, Sowerby, *Min. Con.*, pl. 365.

Occurs at Brickhill, Atherfield, etc.

OSTREA WALKERI, n. sp.

(Plate IV., fig. 4, *a, b.*)

A variable oyster shell—one of the *Rudes*, plano-convex or concavo-convex, covered with rude, imbricating growth-lamellæ: the shells are moderately thick. It is attached sometimes by the whole surface of the lower valve but in other specimens by only a very small area; the lower valve in the former case has a thickened margin. Contour ovate more or less irregular, beak pointed.

Lower valve moderately deep, chamoid, or shallow. Upper valve flat or concave, and irregular with concentric growth striæ or imbrications, which are usually less prominent than on the lower valve.

Ligament area large and triangular (about ⅓th length of the whole shell), its base usually broadest. A median gutter serves to divide it into three sub-equal parts; in the upper valve the ligament area is much smaller.

The principal muscular impression is of moderate size, situated near the anterior side of the shell at about the middle of its length. This species is distinguished from *O. Germanii* (Coquand), and *O. Leymerii* by the inequality and dissimilarity of its valves and by its ligament area. *O. Leymerii* usually has radiating ribs on its lower valve.

One specimen in the Woodwardian Museum which I refer to this species is indistinguishable from *O. deltoidea* (Sow.), and is also very similar to some specimens of *O. Germanii* (Coquand), see Pictet and Campiche, St Croix, pl. 189, f. 2.

Measurements. Length, 1¼ inches; breadth, 1⅛ inches; depth, ⅝ inch; length of ligament area, ⅜ inch; breadth of ligament area, ¼ inch.

Locality. Upware.

N. Europe. Brunswick, Speeton Thon (Coll. Herr von Strombeck).

OSTREA (GRYPHEA) VESICULOSA, Sowerby.

Gryphea vesiculosa, Sby., 1823. *Min. Con.,* pl. 369.
Ostrea vesiculosa, 1869. Coquand, *Monog. des Huîtres,* p. 152, pl. 59, f. 4—7.
„ „ Pictet and Campiche, St Croix, p. 311, pl. 194, f. 1—6.

An upper cretaceous species, but ranging from our Wicken beds (Neocomian) to the Gault and Lower chalk.

Localities. Brickhill. I. of Wight; Warminster, Upper Greensand.

S. Europe. St Croix, Switzerland (Gault).

PECTEN RAULINIANUS, d'Orbigny.

Pecten Raulinianus, d'Orb., 1846. *Pal. Fr., Terr. Crét.* III., p. 595, pl. 433, f. 6—9.
„ „ Pictet and Campiche, *Pal. Suisse, Foss. de St Croix,* p. 202, pl. 172, f. 5—7.

The cross imbrications are well developed in the Upware specimens, traversing the interradial spaces. In some specimens the ribs become double and triple towards the border by the development of side ridges.

Range. Neocomian—Chloritic Marl.

Localities. Upware, Atherfield, Folkestone (Gault), Cambridge Greensand.

S. *Europe.* Presta (Aptien), P. and C.; Gault of departments de l'Aube and de la Meuse.

PECTEN ELONGATUS, Lamarck.

Pecten elongatus, d'Orb., *Pal. Fr., Terr. Crét.,* pl. 436, f. 1—4.

Locality. Brickhill.

N. Europe. Dresden, (Planerkalk), etc.

S. *Europe.* France, Belgium.

PECTEN DUTEMPLEI, d'Orbigny.

Pecten Dutemplei, d'Orbigny, 1845, *Terr. Crét.,* p. 596, pl. 433, f. 10.

„ *Aptiensis,* Pictet and Roux, 1853.

„ *Dutemplei,* Pictet and Campiche, *Pal. Suisse, Terr. Crét.* St Croix, p. 199, pl. 172, f. 1—4.

Distinguished from the other species of the *Interstriatus* group by the nature of the ornament on the buccal ear of the larger valve.

Localities. Upware, Potton, Farringdon, Atherfield ?, Speeton (?).

S. *Europe.* Perte du Rhone (Aptien), d'Ervy (Gault).

Another Pecten from Upware is distinguished by its numerous ribs—over 100. Its ornament is of like nature to that of d'Orbigny's *P. Robinaldinus (Terr. Crét.,* pl. 431, f. 4), but more delicate. It approaches the "type à côtes nombreuses" of Pictet and Campiche from the Urgonien, but that form has only about 80 ribs. Our specimen is too imperfect for satisfactory determination or description.

106   INVERTEBRATA.

PECTEN ORBICULARIS, Sowerby, *var.* MAGNUS, *var. nov.*
(Plate v., fig. 1.)

P. orbicularis, Sby., *Min. Con.,* f. 186.
„   d'Orb., *Terr. Crét.,* pl. 433, f. 14—16.
P. Cottaldinus, Cat. Foss. Mus. Practical Geology, 1878, p. 13
(non d'Orb.).

A shell variously named in English and German collections
P. cinctus, crassitesta, junr. *orbicularis* and *Cottaldinus.* I cannot
distinguish it from the true *Pecten orbicularis* of the Upper Greensand (Warminster, &c.).

P. cinctus (crassitesta) has a much thicker shell with straight
hinge line, the ears being less prominent than in *orbicularis,* and
the shell is marked on one or both valves with delicate radiating
depressed striæ.

Pecten Cottaldinus is quite a different shell, having very unequal
ears, and also ornamented with radiating striæ (see d'Orb., *Terr.
Crét.,* p. 590, pl. 431).

It is worth notice that *Pecten cinctus* is absent from the Wicken
Beds.

Our shell is considerably larger than the original type and deserves the separate varietal name *magnus.*

Pecten orbicularis (Sow.), var. *magnus* nob.

Localities. Upware, Potton, Ely (Neocomian), Hythe, Atherfield, Tealby.

S. Europe. France, Switzerland, &c.

PECTEN (NEITHEA) MORRISII, Pictet and Renevier.

Pecten quinquecostatus, J. Sowerby, *Geol. Trans.,* 1836, Roemer,
Forbes and others [non *P. quinquecos-
tatus,* Sby., *Min. Con.*].
Janira Morrisii, Pictet and Renevier, 1858. *Pal. Suisse, Etage
Aptien,* St Croix, p. 128, pl. 19, f. 2.
„   „   Pictet and Campiche, *Terr. Crét.,* St Croix,
p. 244.

A shell very near to *Neithea quinquecostata* (Sby.), distinguished
from it principally by the character of the areas external to the

outer ribs which are here nearly smooth, while in Sowerby's Upper Greensand shell this part is well costated.

Localities. Upware, Atherfield, Shanklin, Hythe.

S. Europe. Perte du Rhone, Wassy, Spain.

Range. Neocomian.

PECTEN (NEITHEA) ATAVA (Roemer), d'Orbigny.

(Plate IV., fig. 6.)

Janira atava, d'Orb., *Terr. Crét.* IIL p. 627, pl. 442, f. 1—3.

„ „ Pictet and Campiche, St Croix, p. 237, pl. 180.

I only know of one specimen from our neighbourhood and this is in the cabinet of Mr J. F. Walker, M.A.

Locality. Upware.

N. Europe. Hils formation of Shöppenstedt, Berklingen.

S. Europe. Neocomian of France, Switzerland, Spain.

PECTEN (NEITHEA) ORNITHOPUS, sp. nov.

(Plate IV., fig. 5, *a*, *b*.)

Shell subtrigonal, slightly oblique from the greater development of the posterior side, very inequivalve, plano-convex or concavo-convex, auriculate. It is ornamented with six prominent radiating ribs, and subordinate striæ, the former being produced in the form of claw-like processes. The whole surface is crossed with delicate squamous growth-lines.

Lower valve convex with prominent, narrow beak; primary ribs (six) all well defined, the anterior one shortest and the fourth longest; they are angular, and covered with imbricating cross lines.

Interradial grooves slightly narrower than the ribs, depressed, furnished with a number of secondary ribs which are simple and convex, separated by narrow grooves. Fifteen of these are seen in the central furrow, counting those which extend half-way up the sides of the adjacent ribs.

Upper valve concave, plane, or even slightly convex; its primary ribs similar to those of the lower valve, but slightly more rounded; grooves deep, ornamented with radiating ribs and striæ, which are either broad and flat, or numerous and rounded. Ears very

unequal, anterior ear trigonal, slightly truncated, convex in the
right and plane in the left valve; ornamented with the general
imbricating striæ but devoid of rays. Posterior ear very small.
Measurements. Length, 30 mm.; breadth, 25 mm.; thickness,
12 mm.; apicial angle, 55°.

This shell is very generally known from Farringdon and Shanklin as *Neithea Neocomiensis* (d'Orb.). It is however not that
species, our shell being more elongated and more oblique, while the
characters of the intercostal grooves contrast clearly in both valves.
The ears also in d'Orbigny's species are subequal.

I am by no means certain that this is not the original *atavus*
of Roemer. The description (*Nordd. Ool. Sup.*, p. 29) tallies very
well with our shell, but the ears as figured are quite different and
agree with the French *P. atavus.* Yet it is our shell which occurs
'nicht selten' at Schöppenstedt. But unless the original figured
specimen can be discovered, the current ideas (d'Orbigny) are best
to be maintained.

Besides the striking difference of the ears in our species and
P. atavus (d'Orb.), the broader character of the secondary ribs
serves also to distinguish it.

Localities. Upware, Farringdon, Godalming (?).

N. Europe. Schöppenstedt.

HINNITES LEYMERII, Deshayes.

Hinnites Leymerii, Desh., 1842. *Mém. Soc. Geol. Tr.,* tom. v.
　　　　　　　p. 27, pl. 14, f. 1.
　　„　　　„　　Pictet and Campiche, *Pal. Suisse, Terr.*
　　　　　　　Crét., St Croix, p. 224, pl. 175.

A free valve from Brickhill. The internal cast shews well the
intermediate striæ between the principal ribs. Where the shell is
preserved traces of the prominent scales on the ribs are seen.

Locality. Brickhill.

S. Europe. Perte du Rhone.

AVICULA CORNUELIANA, d'Orbigny.
(Plate V., fig. 2.)

Avicula macroptera, Roem., 1841. *Nordd. Ool.* Vol. I. p. 86, pl. 4, f. 5 (non Lamk.).

„ *Cornueliana,* d'Orbigny, *Pal. Fr., Terr. Crét.,* III., p. 471, pl. 389, f. 3, 4.

„ „ *Auctorum.*

„ „ Pictet and Campiche, *Pal. Suisse,* St Croix, p. 66, pl. 152, f. 1—4.

Our Brickhill shell belongs to a round-ribbed type, identical with the one which is abundant in the Schöppenstedt Hils Conglomerat (Brunswick), recognised as the one originally figured by Roemer.

The French type with sharp and more prominent primary ribs (see figs. by d'Orb. and P. and C.) occurs at Speeton, and in the N. German Neocomian clays. These are varietal differences dependent upon habitat.

The shell is commonly called *A. macroptera* in Germany, and is sometimes named *A. pectinata* (Sowerby) in English Cabinets.

The latter is generally considered a different species, distinguished by its smaller sharp ribs, with *one* intermediate rib between each pair of primary ones, while *A. Cornueliana* has a number of riblets and striæ.

Localities. Brickhill, Tealby, Speeton, Folkestone Gault.

N. Europe. Schöppenstedt, Hils Conglom., Bredenbeck.

S. Europe. St Croix (Mid. Neocomian).

PERNA, sp. nov.
(Plate V., fig. 3, *a, b.*)

In form similar to the young of *Perna (Gervillia) aliformis* (Sby.), but the Upware shell is more oblique, its anterior ear is larger, and no distinct ribs are seen.

A shell from Shanklin in the Collection of Mr C. J. A. Meÿer is very near to this, occurring pretty commonly there where mature *aliformis* is not found. (Meÿer.)

Locality. Upware (Coll. Mr J. F. Walker, M.A.).

N. Europe. Oberg, Hildesheim (?), Hils Conglomerat.

PINNA ROBINALDINA, d'Orbigny.

(?) *Pinna gracilis*, Phillips, 1828.   *Geol. Yorkshire*, pl. 2, f. 22.
P. *rugosa*, Roemer (non Schl.) 1839.   *Ool. Geb. Sup.* p. 32, pl.
         18, f. 1—3.
P. *Robinaldina*, d'Orb., *Terr. Crét.* III. p. 251, pl. 330, f.
         1—3.
„   „   Pictet and Renevier, *Pal. Suisse, Terr. Apt.*
               p. 117, pl. 16, f. 5.

Localities.   Brickhill (Woodw. Mus., and in Mr J. F. Walker's ‹
               Cabinet), Atherfield, Speeton (?).
N. *Europe.*   Hildesheim (Göttingen Museum).
S. *Europe.*   Perte du Rhone (Mid. Neocomian).

PLICATULA CARTERONI, d'Orbigny.

(Plate V., fig. 4, *a, b.*)

Plicatula Carteroniana, d'Orb., 1846.   *Pal. Fr., Terr. Crét.* III.
               p. 180, pl. 462, f. 5—7.
„   „   Pictet and Campiche, 1870. *Pal. Suisse,*
               Terr. Crét. St Croix, p. 265, pl. 183,
               f. 3, 4.

A common shell at Upware. The lower valve is usually deeper
than in the French specimen, and the smaller valve is rarely spinous,
but has 8—10 broad, slightly arched ribs. Specimens which have
been found in protected situations (as within oyster-shells) have
the spines on the ribs of the lower valve much produced so as to
form delicate curved spines $\frac{1}{16}$th inch long.

At Brickhill this species occurs as a small variety, with only
five or six (sometimes eight) strong ribs on the lower valve.

Localities.   Upware, Brickhill, Potton, Hythe (Woodw. Mus.),
               Folkestone (Coll. Mr Meÿer).
N. *Europe.*   Brunswick.
S. *Europe.*   Mid. and Lower Neocomian, St Croix, Hau-
               terive, Neuchatel, Bonvillers.

PLICATULA EQUICOSTATA, n. sp.

(Plate V., fig. 5, a, b, c.)

Shell obliquely ovate, plano-convex, very inequivalvular and inequilateral; covered with numerous equal-sized ribs. Upper valve flat or slightly concave, marked with concentric lines of growth, which may be raised into lamellæ especially towards the margin, and ornamented with 15—20 radiating ribs. These ribs are broad and slightly arched; they are separated by only a narrow groove. There are indications that spines may be developed in some specimens at points where the lines of growth cross the ribs. Towards the margin of the shell additional and sharper ribs may be added.

The lower valve is deep and gibbous, attached by a small area at its umbo. It is provided with irregular imbricating growth-lamellæ and zoned by a few, faint or strong varices. These structures are covered with numerous ribs (about 50) which are rounded and gently arched, separated by narrow and simple interspaces crossed by lines of growth.

In one specimen, whose shell has been partly denuded, the ribs on the small valve appear to have been much more numerous than in our type.

Affinities and differences. This species approaches most nearly to *P. placunea* (Lamk.) and *P. imbricata* (Koch.). From the former it is distinguished by its gibbosity and the equality of its ribs, and from the latter by its flat or concave upper valve.

Locality. Upware.

LIMA TOMBECKIANA, d'Orbigny.

L. semisulcata, Forbes, 1845. *Quart. Jour. Geol. Soc.* p. 248 (non Nilsson).

L. Tombeckiana, d'Orbigny, 1845. *Pal. Fr. Terr. Crét.* III. p. 534, pl. 415, f. 13—17.

" " Pictet and Campiche, *Pal. Suisse*, Vol. V., *Terr. Crét.* St Croix, p. 148.

Distinguished from *L. Dupiniana* by its broader, closer-set ribs.

Localities.   Brickhill, Reigate, Tealby.

N. Europe.   In Hils Conglomerat.

S. Europe.   St Croix (Neocomian), very general in the Neocomian of France.

LIMA FARRINGDONENSIS, Sharpe.

(Plate v., fig. 12, *a*, *b*.)

Lima Farringdonensis, Sharpe, *Quart. Jour. Geol. Soc.*, Vol.
x. pl. 5, fig. 2.

A species characterised by its intermediate ribs and by the
striæ on the sides of the ribs, which gradually increase in strength
as the ribs dwindle towards the posterior side.

Localities.   Upware, Brickhill, Farringdon, Atherfield (Leck-
enby Collection).

LIMA LONGA, Roemer.

L. elongata, Roemer, 1836 (non Sby.).

L. longa, Roemer, 1841. *Nordd. Kreid.*, p. 57.

L. undata, d'Orb., 1845. *Pal. Fr.*, *Terr. Crét.* p. 528, pl. 414,
f. 9—12.

L. longa, d'Orb., 1845. *Pal. Fr.*, *Terr. Crét.* III. p. 529, pl.
414, f. 13—16.

L. longa, Pictet and Campiche, *Pal. Suisse*, St Croix, p. 128,
pl. 161, f. 6, 7.

L. undata, Forbes.

A very variable shell, especially in the nature of the ribs and
their interspaces, as may be well studied in the cabinet of Herr
von Strombeck in Brunswick.   In some the ribs are broad and
flattened, in others narrow and rounded; the interspaces are
usually distinctly pitted, but sometimes simple. Some of its varieties are very near to *L. abrupta.* There is a very broad-ribbed
variety in the cabinet of Mr J. F. Walker, M.A., F.G.S., of York.
Some specimens from Upware have lost the outer layers of shell,
and present an almost smooth surface; the imbricating growth-
laminæ being crossed by the faintest 'ghosts' of the ribbing.

Localities. Upware, Brickhill, Potton, Tealby, Farringdon.

N. Europe. Hils Conglomerat, Schöppenstedt, &c.

S. Europe. St Croix, Haut Jura, Lower and Middle Neocomian.

SPONDYLUS ROEMERI, Deshayes.

Spondylus hystrix, Roemer (non Goldfuss), *Nordd. Kreidegeb.* p. 59.

Spondylus Roemeri of French authors; vide D'Orb. *Pal. Fr. Terr. Crét.* III. p. 655, pl. XLV., f. 1—6.

„ „ Pictet and Campiche, *Terr. Crét.*, St Croix, p. 256. De Loriol, Mt. Salève, p. 107, pl. XIV., f. 4, 5.

A very variable shell, having very much the general appearance of the figure of *S. hystrix* Goldfuss (*Pet. Germ.* pl. CV., f. 8), but his description (p. 96), shews it to be perfectly distinct. The spines on the larger ribs vary much in number and in prominence, while some specimens are devoid of spines altogether. The ribs are rather closer set than in the figures by D'Orbigny and De Loriol above referred to.

The lower valve is smaller, flattened and marked with strong growth lamellæ over its broad base ; its upturned border is covered with numerous simple-rounded ribs without spines.

Localities. Brickhill.

N. Europe. Schöppenstedt, Hils Conglomerat.

S. Europe. St Croix and Mt. Salève, Mid. Neocomian ;

Range. Valengien, Urgonien.

TRIGONIA UPWARENSIS, Lycett.

Trigonia spinosa of fossil lists from the Upware deposit. Walker, 'Coprolite Workings in the Fens,' *Geol. Mag.,* 1867, p. 310.

Trigonia Upwarensis, Lycett, Monog. Brit. Foss. Trigoniæ, *Pal. Soc.* p. 143, pl. XXXIX., f. 4.

I believe this species also occurs in the 'derived' blocks of dark irony grit rock at Upware ; but these specimens are not well preserved.

Locality. Upware (not uncommon).

K. 8

114   INVERTEBRATA.

NUCULA SUBTRIANGULA, Koch and Dunker.

? *Nucula subtrigona*, Roemer, 1836.   *Nordd. Ool.* pl. VI., f. 6.
Nucula subtriangula, Koch and Dunker, 1837, *Beitr. Nordd.
Ool.*, p. 50, pl. VI., f. 1.

Localities. Upware, Shanklin.
N. Europe. Hils Thon, Elligserbrink.

ARCA MARULLENSIS, D'Orbigny.

Arca Marullensis, D'Orb., *Pal. Fr. Terr. Crét.* III., p. 205, pl.
          CCCX., f. 3—5.
  „   „   Pictet and Campiche, *Pal. Suisse, Terr.
          Crét.*, St Croix, p. 432, pl. CXXX., f. 1—4.

The ligament area is deeply retreating.

Localities. Upware, Farringdon.
S. Europe. Morteau (Doubs), Urgonien infr.

ARCA CARTERONI, D'Orbigny.

(Plate V., fig. 7, *a, b.*)

Arca Carteroni, D'Orb., *Pal. Fr. Terr. Crét.*, III., pl. CCCIX.,
          f. 4—8.
  „   „   Pictet and Campiche, *Pal. Suisse Terr. Crét.*,
          St Croix, p. 436, pl. CXXX., f. 9.

A rare shell, belonging to the sub-genus *Byssoarca*. The
lozenge-shaped ligament area is even more developed in our specimens than in the figures of French shells; it is crossed by as many
as eight pairs of the angular ligament grooves.

Localities. Upware, Atherfield, Peasmarsh, I.W.
S. Europe. St Croix, Landeron.
Range. Mid. Neocomian, Urgonien.

CUCULLÆA SUBNANA, Pictet and Roux.

(Plate v., fig. 10.)

Arca subnana, Pictet and Roux, 1852, *Moll. Foss. grès verts*, p. 461, pl. XXXVI., f. 6.

„   „   Pictet and Campiche, *Pal. Suisse Terr. Crét.*, St Croix, p. 466.

Near to the common Atherfield species, *C. Cornueliana*, D'Orb., but distinguished by wanting the ridge which always passes obliquely along the middle of the posterior area in that species.

It is distinguished from *C. nana* by its ligament furrows.

Our shell is double the size of the figured types.

Localities. Upware, Farringdon.

S. Europe. Perte du Rhone, Gault; St Croix, Lower Gault.

CUCULLÆA, sp. ?

(Plate v., fig. 8.)

A somewhat similar shell to the last, but more oblique—a cuneiform shell. The umbo is not prominent, hinge rather short, hinge area small. The radiating striations are very delicate, for the most part obscure.

I only know of one specimen (Woodw. Mus.), and this seems to be partly decorticated.

Measurements. Transverse diameter, 25 mm.; depth, 19 mm.; umbo to posterior angle, 21 mm.

Locality. Upware.

PECTUNCULUS SUBLÆVIS, Sby.

(Plate v., fig. 9, *a*, *b*, *c*.)

Pectunculus sublœvis, Sowerby, 1824, *Min. Con.*, pl. CCCCLXXII., f. 4.

The umbo is narrower and less elevated than is usual in this species, and the ribs are broad and flattened.

The ribs vary in this species.

Localities. Upware, Blackdown, Shanklin, I.W.

116 INVERTEBRATA.

PECTUNCULUS MARULLENSIS, Leymerie.

(Plate V., fig. 11.)

Pectunculus Marullensis, Leym. 1842, *Mém. Soc. Géol. France*,
       ` ` v., pl. IX., f. 2.
    „   „   D'Orbigny, *Terr. Crét.*, p. 187, pl.
        CCCVI., f. 1—6.

The ribs vary; they do not shew such a regular cross-line
structure as is shewn in M. D'Orbigny's figures.

Localities. Upware, Shanklin.
S. Europe. Paris basin, Middle Neocomian.

PECTUNCULUS OBLIQUUS, sp. nov.

(Plate VI., fig. 1, *a, b, c, d.*)

Shell thick, ovate-oblong, transverse, slightly oblique; convex,
but somewhat flattened in the centre; surface marked with lines
of growth and ornamented with delicate radiating striæ.

Umbones contiguous, rather depressed, oblique, situated in
about the line of the anterior third of the shell. Hinge line short,
less than half the length of the shell.

Ligament area narrow with several (about eight) angular
furrows. Hinge strong, furnished with few central transverse
teeth, and three longitudinal lateral teeth upon each side of the
semicircular hinge shelf.

The surface striæ are very numerous, gently convex, with
simple, narrow interspaces.

The species is best distinguished by its shape, which is almost
cuneiform, and its anterior, oblique umbones. It approaches nearest
to some Jurassic species from the Great oolite and Coral rag.

Measurements. Antero-posterior diameter, 16 mm.; umbono-
pallial diameter, 13 mm. Thickness (through both valves),
10 mm.

Locality. Upware.

MODIOLA PEDERNALIS, Roemer?

(Plate VI., fig. 2, *a, b*.)

Modiola pedernalis, Roem., *Kreide Texas*, pl. VII., fig. 11.

The ribs are very delicate in our Upware shells, and the cardino-lateral border is straighter and less curved than in Roemer's figure, but having compared it with specimens from the original locality in the Göttingen Museum I believe them to belong to the same species.

Locality. Upware.

N. Europe. Elligserbrink, Hils Thon.

Friedrichsberg, in Texas, America.

MODIOLA OBESA, sp. nov.

(Plate VI., fig. 3, *a, b*.)

Description. A very obese species, elongated, oval, gibbous, and arcuated. Cardinal and cardino-lateral borders forming together one simple curve; pallial border markedly sinuated; anal border narrow and rounded. The umbones are nearly terminal, slender, depressed and incurved; an arched shoulder runs from them to the front limit of the anal border. Buccal area swollen, marked off by a shallow furrow running alongside the mesial shoulder,—its pallial termination corresponding with the sinus, which is situated towards the posterior end, nearly two-thirds of the whole length from the umbo.

The surface is marked only by delicate laminæ of growth, and coarser growth-stages.

Measurements. Length, 21 mm.; breadth, 10 mm.; thickness (through both valves), 13 mm.

Affinities and differences. A shell of the *Modiola æqualis* type, also approaching near to *M. Montmolini* of MM. Pictet and Campiche, *Terr. Crét.*, St Croix (pl. CXXXIII., f. 2); from both of these it is distinguished by the regularly-arched cardinal side, instead of having a distinct cardinal angle.

Our species is more elongated than *M. æqualis*, but less so than *M. Montmolini*. The depth of the shell is a characteristic feature, and in this as well as in the regular curvature of the cardinal side, it shews affinity to *M. lineata*, Sow. (Fitton, pl. XIV., fig. 2); but the absence of striation and the gibbous buccal region readily distinguish it.

Localities. Upware, Shanklin, I.W.

MODIOLA, sp. nov.

An ovate, lanceolate shell, convex, slightly arcuated, covered with delicate, crimp-like striations, which are well developed upon the buccal side, but appear to become fainter over the back. The umbones are terminal, slender, oblique, and incurved. Cardino-lateral angle gentle and rounded. The buccal area is but faintly marked off, smooth, and slightly swollen.

Measurements. Length, 2¼ inches; breadth, 1 inch; thickness (across both valves), 1 inch.

This shell is near to *M. Cottæ*, Geinitz, but the author of that species agrees with me that the two shells are distinct, the umbo being much more obtuse in *M. Cottæ*. The ornament striæ have the character of *M. lineata*, Sow., a shell totally different in shape.

We only know of a single specimen, and in this the shell itself is wanting over most of the back; therefore from being thus imperfectly known I decline to give a name to the species.

Locality. Upware.

CARDIUM COTTALDINUM, D'Orbigny.

(Plate VI., fig. 4.)

Cardium Cottaldinum, D'Orb., *Pal. Fr. Terr. Crét.*, III. p. 22, pl. CCXLII., f. 1—4.

„   „   Pictet and Campiche, *Pal. Suisse Terr. Crét.*, St Croix, p. 146, pl. CXVIII., f. 1, 2.

A larger shell than *C. Ibbetsoni* of Atherfield, and without the posterior angles. Its greater cardino-pallial diameter distinguishes it from *C. Voltzii* Leym. The best specimens are in the cabinet of Mr J. F. Walker, M.A., of Sidney Sussex College, one of which shews the hinge. The teeth are small and slender; cardinal tooth slightly bifid, anterior lateral (of the left valve) laminar, double, posterior lateral obsolete; behind the umbo a well-marked ligament support is seen.

Localities. Upware, Folkestone, Shanklin, Punfield, under the Punfield series (Meÿer).

S. Europe. Middle Neocomian of Landeron and Mont Salève.

Range. Neocomian.

CARDIUM SUBHILLANUM, Leymerie?

Cardium subhillanum (Leym.), 1842, *Mém. Soc. Géol. Fr.* v., p. 5, pl. VII., f. 2.

„   „   De Loriol, 1861. *Descr. Anim. Invert. du Mont Salève*, p. 81, pl. X., f. 4.

I only know of one specimen, and in this the shell itself has been lost, leaving only the inside cast. Therefore there is some doubt in this identification. The cast is figured by Pictet and Campiche, *Pal. Suisse Terr. Crét.*, St Croix, pl. CXXI., f. 3, 4.

Localities. Upware, Atherfield, Devizes, Hythe.

S. Europe. St Croix, Landeron, Mt. Salève, Jura.

CYPRICARDIA STRIATA, Geinitz, sp.

Cardita striata, Geinitz, *Charak.* II. p. 52, pl. X., f. 3 a—c.

Modiola carditoides, Geinitz, *Das Elbthalgebirge in Sachsen* I. p. 218, pl. XLVIII., f. 11—13, pl. XLIX., f. 19, 20, in Dunker and Zittel, *Palæontographica*.

The discovery of the hinge teeth in the closely-allied species *C. squamosa*, n. sp. (see *infra*) proves this to be a *Cypricardia*. I have no doubt as to the above identification of this shell, having compared specimens with Professor Geinitz in the Dresden Museum, but it is remarkable that this little shell should occur only in two such widely separated localities, and in rocks of such different ages.

120 INVERTEBRATA.

The muscle markings and pallial line are shewn upon an internal cast from Oberau, Dresden. The two adductor impressions are placed low in the shell; they are rounded, and connected by a simple, nearly straight pallial line.

Localities. Upware.

N. Europe. Lower Quader conglomerates, Oberau, near Dresden; and Lower Pläner, Plauen, Dresden (Upper Cretaceous).

CYPRICARDIA SQUAMOSA, sp. nov.

(Plate VI., fig. 5, *a—c.*)

Shell quadrate, but broader behind, gibbous, ornamented with squamous growth laminæ, crossed by more or less distinct radiating striæ. Umbo quite anterior, prominent, incurved; a strong rounded shoulder passes from the umbo obliquely across the shell. Anal border rounded; buccal region slightly swollen; a gentle depression meets the pallial border in front of the median line. Lunule cordiform, deeply impressed.

In the hinge two obtuse prominent teeth are seen in the left valve. Right valve with (?)one cardinal tooth; no laterals.

The ornament consists of a few (seven or eight) prominent, separate concentric lamellæ; and radiating striæ which are usually faint, best marked over the back, and disappearing on the posterior side.

Measurements. Antero-posterior diameter, 13 mm.; dorso-pallial diam., anteriorly, 9 mm., posteriorly, 11 mm.; thickness (through both valves), 10 mm.

Nearly allied to the *Cardita Neocomiensis* of D'Orbigny, but less gibbous in shape, and further distinguished by its few, prominent squamous lines of growth, and the faint development of the radiating ribs.

Locality. Upware.

CYPRICARDIA ARCADIFORMIS, sp. nov.

(Plate VI., fig. 6.)

Shell oblong, *Arca*-shaped; the sides straight, posterior border oblique. Pallial border sinuated. A prominent angular shoulder runs from the umbo diagonally across the shell. Umbones promincnt, incurved, quite anterior. Buccal region slightly swollen. Lunule cordiform, impressed; a posterior lanceolate lunette is also present. The surface is cancellated with prominent growth laminæ, crossed by costæ radiating out from the umbo.

The posterior region behind the shoulder is smooth or with faint shadowy costæ.

Measurements. Antero-posterior diameter, 21 mm.; dorso-pallial diameter, 12 mm.; thickness (through both valves), 13 mm.

Locality. Upware.

CARDITA ROTUNDATA, Pictet and Roux?

(Plate VI., fig. 7).

Cardita rotundata, Pictet and Roux. *Moll. Foss. Genève,* p. 443, pl. XXXIII., f. 6.

A lower gault species occurring at Folkestone, where it is known by this name; but it is a longer shell (length equal to breadth) than the Swiss type, and its ribs are fewer and stronger.

Localities. Upware, Folkestone, Potton.

S. Europe. Depts. Aube, Yonne, Ardennes, Meuse.

Range. Neocomian (Upware) and Lower Gault.

OPIS NEOCOMIENSIS, D'Orbigny.

(Plate VI., fig. 8, *a—c.*)

Opis Neocomiensis, D'Orbigny, *Pal. Fr. Terr. Crét.,* III., p. 51, pl. CCLIII., figs. 1—5.

Opis Desori, De Loriol, *Foss. du Mont. Salève,* p. 66, pl. VIII., f. 4.

The shell varies slightly in breadth. There can be little doubt that MM. Pictet and Campiche are justified in their conclusion that *O. Desori* (De Loriol) is the same species as the *O. Neocomiensis* of D'Orbigny. The differences between them are slight, and, according to Pictet and Campiche, depend upon differences in the ages of the specimens.

Our Upware shells are sometimes corrugated with projecting growth zones.

Specimens in the Woodwardian Museum shew the teeth. There are in the right valve one large tooth, V-shaped and massive, by whose sides are two sockets, which are bounded externally by tooth-like ridges. In the left valve are two diverging teeth, one of them corresponding with the margin of the wide, open lunule. The other is supported on the cardinal shelf.

Localities. Upware, Coleshill, Farringdon.

N. Europe. Elligserbrink, Brunswick (Hildesheim Museum).

S. Europe. Hauterive, Mt. Salève, Landeron.

Range. Valengien, Neocomian.

ASTARTE, sp. nov.

(Plate VI., fig. 9.)

A shell approaching the *Astarte Valengiensis* of Pictet and Campiche, but more orbicular.

Also *A. Valengiensis* is broader on the pallial side, and narrower and more acuminated on the cardinal side than our shell.

Localities. Upware, Potton, Tealby ? Shanklin.

ASTARTE SUBDENTATA, Roemer.

(Plate VI., fig. 11.)

Astarte subdentata, Roemer, 1841. *Nord. Deutsch. Kreid.,* p. 71, pl. IX., f. 9.

Localities. Upware, Potton.

N. Europe. Hils Thon. (Hildesheim Museum).

ASTARTE, sp.

(Plate VI., fig. 10. *a, b.*)

A more globose shell than *A. subdentata,* Roemer ; subtriangular and rounded in contour ; length and breadth nearly equal. It is nearly equilateral, the posterior side being slightly the longer. The umbones are large and prominent. The surface is marked by delicate concentric lines, and also usually by more conspicuous growth-grooves or ridges, but these are not prominent and regular.

Anterior side rounded, posterior slightly truncated.
The lunule and lunette are shallow and ill-defined.

Measurements. Antero-posterior diameter, 12 m.m.; umbono-pallial diameter, 11 mm.; thickness (across both valves), 8 mm.

Locality. Upware.

CYPRINA SEDGWICKII, Walker, sp.

(Plate VI., fig. 12, *a, b, c.*)

Sphæra Sedgwickii, Walker, *An. and Mag. Nat. Hist.*, November, 1866, pl. XIII., figs. 1, 2.

Cyprina angulata, var. Seeley, *An. and Mag. of Nat. Hist.*, 1867 (non Flem., non Sby.)

Cyprina Sedgwickii, Keeping. List of Fossils in Bonney, 'Cambridgeshire Geology,' Appendix, p. 68.

Venus   *Catalog. Cret. Foss., Museum of Practical Geology,* 1878, p. 21.

I append Mr Walker's description :

" Shell globose, nearly equilateral, slightly gibbous, concentrically furrowed, the striæ finer and more distinct towards the ventral margin; ligament prominent; lunule large, distinct, cordate."

Measurements. Length, 1·6 in.; breadth, 1·5 in.; thickness, 1·3 in.

The size is very constant; besides the delicate growth striæ, the shell is usually zoned with gentle varices. There is a faint posterior ridge running from the umbo to the postero-lateral margin, in which the shell approaches the character of *C. angulata.*

In outline and in thickness the species varies somewhat, the length and breadth being in some specimens equal. Such forms are also more tumid, and come to resemble *C. regularis* (D'Orbigny) of the gault. But the two species are distinct.

Since Mr Walker described this shell as *Sphæra Sedgwickii*, the hinge has been worked out in a number of specimens. There are three cardinal teeth, the anterior one being under the lunule ; this tooth is double in the right valve; there is a distinct posterior lateral tooth. The pallial line is simple. Upon the inside cast the adductor muscle-marks are well seen, and the posterior ridge is more distinct than upon the shell itself. The umbo is pointed. Pallial line simple.

Our shell has considerable resemblance to some forms of *C. angulata, rostrata* (Sby.); their hinges are similar; but *C. Sedgwickii* is a smaller species, with a more globose shell, and less expanded in front. The posterior angle is not nearly so well developed, though it is faintly seen.

The distinct cordiform lunule, marked off by a definite groove, is an important character, serving to separate off a small group of cretaceous shells from the true *Cyprina*. It is seen in *C. regularis*, as described and figured by D'Orbigny, and this author states that the same structure occurs in the shell which he refers to *C. rostrata*, Fitton; I have not detected it in British specimens of that species.

Localities. Upware, Potton.

N. Europe. Hils Thon (in Hildesheim Museum as "*Mactromya nicht bestimmt.*")

CYPRINA SEDGWICKII, Walker, var.

A rounded, inflated variety of *C. Sedgwickii*, Walker, approaching the gault species *C. regularis*.

CYPRINA ANGULATA, var. ROSTRATA, Sby.

Cyprina angulata, Sowerby, *Mineral Conchology*, pl. 65, var.
Cyprina rostrata, Sowerby, *Geol. Transactions*, vol. IV., p. 341, pl. 17, f. 1.
C. Ervyensis, Pictet and Roux.

We only know of one specimen, and that an internal cast, of this species from Upware. It agrees with Sowerby's typical *C. rostrata* from Blackdown.

Localities. Upware, Potton (Coll. J. F. Walker), Blackdown, Sandgate, Hythe.

N. Europe. Perte du Rhone, Saxonet, Mortran.

Range. Lower Greensand, Aptien, and Lower Gault.

CYPRINA OBTUSA, n. sp.
(Plate VI., fig. 13, a—c.)

Shell subtriangular, inflated, very inequilateral, the surface marked with delicate concentric striæ. Umbones prominent and pointed, directed forwards and incurved. Anterior side rounded; posterior side slightly produced, its margin rounded; antero-cardinal region hollowed, but with no defined lumule. Postero-cardinal area well-defined by a strong keel, which arches backwards from beneath the umbo to the postero-lateral margin, so as to mark off the posterior lunette. Behind this keel comes a second oblique ridge, which is however very feebly developed; it produces a slight angle in the pallial border of the shell, which is elsewhere simply curved.

Affinities and differences. *Cyprina cuneata, C. Saussuri* (Brong.) and *C. obtusa* all belong to one group of *Cyprina* characteristic of the cretaceous rocks, and distinguished by their wedge-shape, produced posterior side, and the presence of two posterior oblique ridges. They are closely allied to one another, and may, I believe, prove to be only varieties of one species. But materials are not yet collected to prove this, though the idea is supported by some Atherfield specimens.

As compared with *Cyprina cuneata*, Sby., the umbo is less prominent and more rounded, all the angles less sharp and the posterior side less produced, so that the pallial border is scarcely, if at all, sinuated. The whole shell is less angular and more tumid, hence the name applied to it. In many of these differences our shell approaches *C. Saussuri* (Brong.), but that shell is of a different shape—not so trigonal—also its umbo seems to be more prominent, and the posterior end is more produced.

In general shape it is near to *C. Bernensis* Leym. (l'Aube, pl. v., f. 5).

Localities. Upware, Folkestone, base of Folkestone Beds (Mr Meÿer's collection).

VENUS VECTENSIS, Forbes.

Venus Vectensis, Forbes, 1845. *Quart. Journ. Geol. Soc.,* Vol. I., p. 240, pl. II., f. 4.

So far as I know only one specimen has occurred, and that a very indifferent one (Woodw. Mus.)

A specimen in the University collection from Atherfield shews the hinge teeth in the right valve. There are three cardinal teeth and a fourth tooth, which starts under the umbo, runs obliquely along the posterior cardinal margin.

Localities. Upware, Atherfield, I.W.

PANOPÆA PLICATA, Sowerby.

Mya plicata, Sowerby. *Min. Con.* pl. CCCCXIX., f. 3.

Localities. Upware, Atherfield, Sandown, Speeton, Sandgate, Hythe.

N. Europe. Bredenbeck, Hils Thon.

S. Europe. Perte du Rhone, St Croix, Vassy.

PANOPÆA GURGITIS, D'Orbigny.

Panopæa plicata, Goldfuss., *Pet. Germ.,* p. 274, pl. CLVIII., f. 5 (non Sby.).

Panopæa gurgitis, D'Orb., *Pal. Fr. Terr. Crét.,* III., p. 345, pl. CCCLXI., f. 1, 2,

Localities. Upware (Coll. Mr J. F. Walker).

S. Europe. Dept. Var. (Cenoman.)

? THRACIA or TELLINA.

(Plate VI., fig. 14).

A doubtful shell, of which only two specimens have been discovered at Upware. It bears much resemblance to the figures of *Thracia Couloni,* Pictet and Renevier, *Pal. Suisse, Terr. Aptien,* p. 66, pl. VII., f. 4.

Locality. Upware.

BORING SHELLS.

Lithophagous bivalves lived in great abundance at the time when the Ironsand and Phosphatic Series was being deposited, so that commonly the phosphatic nodules are studded with the openings of their small burrows, which are usually about so big as to admit easily a large pin's head. Some shells collected from these burrows shew that they were made by small *Modiola* and *Arca*-like bivalves (*Lithodomus* and ? *Saxicava*).

Larger pyriform crypts, measuring 6—8 mm. deep are also occasionally found, some of them in fossil wood, and these probably belong to the *Pholas constricta* of Phillips.

Another and larger species is evidenced by a series of fine clavate crypts in the paddle-bone of a *Pliosaurus*, which measure 4 inches in depth, 1 inch in diameter. Their courses are somewhat tortuous. I have found similar great burrows in the Hythe beds of Hythe, Kent.

One very curious organism, probably of the same nature, was found at Upware :—a pyriform shelly structure, with small circular opening, is continued below into an irregular swollen tube, which, in one specimen, turns sharply upwards and, uniting with a fellow tube, ends in a large, slightly expanded, irregular common opening. Altogether the specimen has much resemblance to a Scotch bagpipe.

PHOLAS (FISTULANA) CONSTRICTA, Phillips.

Pholas constricta, Phillips, *Geol. Yorkshire*, pl. II., f. 17.
Roemer, *Nordd. Kreide*, pl. x., f. 11.

Localities. Upware, Speeton.

LITHODOMUS PHOSPHATICUS, sp. nov.

(Plate VII., fig. 1, *a, b.*)

A small, rather variable boring shell, abundantly found in crypts in the phosphatised casts of Ammonites, &c.

Description. Shell small, cylindrical, the surface smooth or marked only by lines of growth. *Umbones* nearly terminal, broad, flattened, and incurved. Two gentle shoulder ridges start from the umbo, the one more obtuse, running to the anterior pallial margin, and the other, which is angular in front, passing obliquely to the posterior end; a slight sinus is therefore formed in the pallial border.

The casts shew a posterior ligament groove, bordered by ridges.

Measurements. Length, 7 mm.; breadth, 3 mm.; thickness (through both valves) 3 mm.

Localities. Upware, Potton, Brickhill.

MODIOLA ARCADIFORMIS, n. sp.

(Plate VII., fig. 2, *a, b.*)

Like the preceding this is a small boring shell found in the phosphatic masses.

Shell small, oblong, convex (*Arca*-shaped); anterior side rounded; pallial border gently convex; anal border square-truncate.

The *umbo* is large, rather prominent, but flattened, incurved and pointed; it is well removed from the anterior end. A sharply angular keel runs from the umbo to the posterior pallial margin, cutting off a distinct, concave posterior region. There is a strong internal ligament ridge.

The surface is covered with distinct lines of growth, which are sharply angulated over the keel. The species is well distinguished by its *arca*-like shape, the posterior keel and hollowed posterior area.

Localities. Upware, Potton.

BRACHIOPODA.

(See *ante*, p. 20.)

MEGERLIA (KINGENA) RHOMBOIDALIS, n. sp.

(Plate VII., fig. 3, *a—c.*)

Description. Shell rhomboidal, slightly longer than broad, moderately convex, thickest near the beaks. Surface smooth or marked with lines of growth. Shell structure coarsely punctate.

Dorsal valve obtusely trigonal, convex, especially near the beak. Hinge line gently curved; front obtusely pointed. A distinct line extending $\frac{3}{4}$ the length of the shell indicates the internal dorsal septum of the genus.

Ventral valve rhomboidal, obtusely keeled from back to front. Beak sharp; beak-ridges acute, area elongated, flattened and concave.

Deltidium in two pieces; foramen small.

Measurements. Length, $9\frac{1}{2}$ mm.; breadth, 9 mm.; thickness, 6 mm.

Affinities and differences. Characterised by its rhomboidal form. In Britain we have only *M. lima,* Defr. to distinguish it from, a species to which it bears considerable resemblance; but the front margin in *M. lima* is truncated, whilst in *M. rhomboidalis* it is obtusely pointed; in our new species the beak is less incurved, the hinge line is, as a rule, more curved, and the dorsal valve is more swollen than in *M. lima;* also this latter species is ornamented with "small granulations or short, hollowed spines" (p.43, Davidson), which I have not found in *M. rhomboidalis.* Again, the lines of growth are more marked.

In geological times the two species seem always to have been separated, *M. lima* ranging only from the Upper Gault to the Chalk.

Locality. Brickhill.

WALDHEIMIA BONNEYI, sp. nov.

(Plate VII. fig. 4, *a, b, c.*)

Shell thick, oval-pentagonal, usually inflated; pointed behind, truncated and straight or sinuated in front. It is broadest near the centre, and tapers behind to the beak. The surface is marked with interrupting lines of growth of varying intensity, valves about equally convex, deepest near the centre.

Ventral valve moderately convex, impressed towards the front so as to form the lateral ridges, and gently carinated in the posterior region. Beak rather prominent, slightly incurved, beak ridges sharp.

The hinge area is large and flattened. *Deltidium* large, in two pieces. *Foramen* small.

The *Dorsal valve* is pentagonal, somewhat inflated, the sides gently curved or with a slight angle. Front sinuated. The *loop* is long and reflected.

This species varies considerably in its form much in the same way as does its near ally *W. pseudojurensis.* Some of the specimens are flattened so as to pass into that species by regular gradation, but the typical ones are always more inflated. The less truncated forms approach *W. Wanklyni.* Some specimens are much thickened on the front margin, some are broadly truncated in front, and others are almost pointed.

K.   9

Measurements. Length, 33 mm.; breadth, 26 mm.; thickness, 18 mm.

Another specimen gives length 30 mm.; breadth, 23 mm.; thickness, 20 mm.

TEREBRATELLA KEEPINGII, Walker.

(Plate VII. fig. 19.)

Shell elongated, subpentagonal, or sometimes trigonal; posterior end produced. Front margin truncated, forming distinct lateral angles.

The whole shell is covered with a beautiful delicate striation radiating from the beaks, new striæ or ribs being frequently intercalated towards the front. Striæ rounded, about equal to the interspaces in size.

Dorsal valve moderately convex, flattened along the middle and depressed in front. Dorsal septum elongated.

The *ventral valve* is convex with steep sides, straight in front, strongly arched across behind, and produced into a very prominent long beak. Beak ridges well defined and sharp near the foramen. Area elongated, slightly convex. *Deltidium* elongated in two pieces. *Foramen* small.

This shell varies considerably: the typical shell is pentagonal in form, but some specimens have the long sac-like shape of *T. digona;* they vary also in the length of the shell and coarseness of striation; the latter is however always much more delicate than in *T. Davidsoni.*

The last-named species is the nearest ally of *T. Keepingii,* and is the only one with which it could be confounded. Our new species is distinguished by its greater length, especially length of beak and area, its steep sides, truncated front, and delicate striæ. It is an extremely well-marked type, although as I have shewn elsewhere the passage forms have been found connecting it with *T. Davidsoni* (see *ante,* p. 25).

Measurements. Length, 23 mm.; greatest breadth, 17 mm.; thickness, 11 mm.

ANNELIDA.

(See *ante*, p. 28.)

SERPULA LOPHIODA, Goldfuss.

(Plate VII. fig. 5, *a*, *b*.)

Serpula lophioda, Guldfuss, *Petrifacta Germaniæ*, I., p. 236, pl. 70, fig. 2.

A fine specimen, five inches long, in the University collection.

Locality. Upware.

N. Europe. Essen, Shöppenstedt.

SERPULA RUSTICA, Sowerby.

(Plate VII. fig. 6, *a*, *b*.)

Serpula rustica, Sby., *Min. Con.*, pl. 599, f. 3.

An obtusely quadrangular tube, commonly bent into a loop in the Upware specimen, the looped portion being attached, but the rest of the tube remaining free and straight. *Serpula quadrangularis* (Roemer) is its representative in the German Neocomian (Hils formation).

Localities. Upware, Brickhill, Godalming (in Lower Greensand), Folkestone (Upper Greensand).

SERPULA ARTICULATA, Sowerby.

(Plate VII. fig. 7.)

Serpula articulata, Sby., *Min. Con.*, vol. VI., p. 204, pl. 599, f. 4.

A worm-tube of the *vertebralis* (Oxford clay) type. The aperture expands periodically, trumpet fashion, and the successive enlargements produce the *articulated* appearance of the test. This is a rougher, more irregular worm-tube than *S. vertebralis*, with unequal intervals between the quadrangular nodulose varices, but such differences may be produced simply by difference of habitat. Our *Serpula rustica* from Upware sometimes exhibits very similar structure.

Localities. Upware, Brickhill, Godalming.

N. Europe. Berklingen (Hils Conglomerate).

9—2

SERPULA ANTIQUATA, Sowerby.

Serpula antiquata, Sby., *Min. Con.*, pl. 598, f. 4.

Localities. Upware, Wiltshire (Upper Greensand).
N. Europe. Bredenbeck, Elligserbrink.
S. Europe. Mont Salève.
Range. Neocomian to Lower Chalk.

SERPULA PLEXUS, Sowerby.

Serpula plexus, Sowerby, *Min. Con.*, pl. 598, p. 1.

Localities. Upware, Blackdown, Wiltshire (U. G. S.), Norwich (Chalk).

SERPULA GORDIALIS, Goldfuss.

Serpula gordialis, Goldf. *Pet. Germ.*, pl. 71, f. 4.

Localities. Upware, Brickhill, Farringdon.
N. Europe. Schöppenstedt.
S. Europe. Perte du Rhone.

SERPULA AMPULLACEA, Sowerby.

Serpula (Vermillia) ampullacea, Sby., *Min. Con.*, pl. 597, f. 1—5.

Localities. Brickhill, Upware, Blackdown, Norwich (Upper Chalk).
N. Europe. Schöppenstedt.

SERPULA, sp. ?

Spirorbis Phillipsii, de Loriol, *Foss. du Mont Salève*, p. 154, pl. 22, f. 15.
? Serpula Phillipsii, Roem.

Tube attached by a broad circular base, upon which the tubes coil to form a hollow cylinder like a thick rope-coil. It may be a form of *Vermicularia Phillipsii*, Roemer, but I have never seen the connecting forms.
Localities. Brickhill.
N. Europe. Schöppenstedt.

VERMICULARIA PHILLIPSII, Roemer.

Vermicularia Phillipsii, Roemer, *Kreide*, p. 102, t. 16, f. 1.

,, *Sowerbyi*, Phillips, *Geol. Yorkshire*, I. pl. II., f. 29.

Localities. Upware, Brickhill, Speeton.

N. Europe. Schöppenstedt, Heligoland.

VERMICULARIA POLYGONALIS, Sowerby.

(Plate VII. fig. 8, *a, b.*)

Vermetus polygonalis, *Min. Con.* pl. 596, f. 6.

When the outer layer of the shell is removed the keels on the whorls become very obscure.

Localities. Upware, Hythe; Sandgate.

N. Europe. Near to *V. quinque carinata*, Roemer, from the Bredenbeck Hils thon.

ECHINODERMATA.

(See *ante*, p. 28).

CIDARIS, sp. nov. ?

(Plate VII. fig. 9.)

Several examples of the stout thorny spines of a *Cidaris* are preserved in the Woodwardian Museum.

An imperfect specimen measures $3\frac{1}{2}$ inches in length, the shaft being 4 mm. in diameter at its head end, and tapering to $2\frac{1}{2}$ mm. at its (broken) apex. The thorns are irregularly scattered, and appear to have been directed upwards; both shaft and thorns are rounded in section. Between the thorns the shaft appears to the unaided eye to be smooth, but the magnifying glass shews it to be delicately striated longitudinally. The head and collar are rather small (greatest diam. 5 mm.), and have a tumid and obtuse appearance.

Some occasional and exceptional spines occur amongst some of the undermentioned species of *Cidaris*, which are undistinguishable from the Upware spines. But these latter are the proper type of the species, and not exceptional, as is proved by the constancy of their character in all the known specimens. I have not been able therefore to identify them with any known species.

C. verticillata (Stoppain) differs in that its thorns are grouped in verticals. In *C. spinigera* the spines are more slender, and are usually curved (Cotteau), the thorns also being longer and more spine-like.

It is nearest to *C. spinosa*, Agassiz, of the Jura, Terr. à Chailles at Fringeli, found also in the Jurassic of Normandy and N. Germany. " Un espèce très voisine, mais cependant spécifiquement différente, provenant de l'argile de Specton, est figurée dans Phillips, *Geol. Yorkshire*, t. II., f. 3." This figure shews the thorns arranged, somewhat irregularly, in zones.

The thorny spines of *C. muricata* in the Hils Conglomerat vary much as seen in the Gottingen Museum, but they are distinguished from our Upware species by having the interspaces between the thorns muricated.

Some forms of the *C. variabilis* (Dunker and Koch) agree very well with ours, but the spine is sharper, and we do not find any of the typical forms of this species at Upware. Other thorny spines occur in the Cretaceous of Rugen (Desor), and in the Kimmeridge Clay of Ely (Woodwardian Museum).

A single interambulacral plate in the Woodwardian Museum, from Upware, may have belonged to the same Urchin. It does not agree with Ooster's figure of what he believes is the plate of *C. spinifera*. It is transversely oblong, densely granulated, with large areole, wide and rather prominent boss, with coarsely crenulated summit.

Localities. Upware, Brickhill, Farringdon (?), Godalming.

CIDARIS ? sp.

The pyramid of jaws (alveoli) of a large Urchin, measuring one inch in height and breadth.

Locality. Upware (Coll. Mr J. F. Walker).

CIDARIS FARRINGDONENSIS, Wright?

? *Cidaris Farringdonensis*, Wright, *Cretaceous Echinodermata, Palæontogr. Soc.*, pl. II., f. 6—8.

Only a single spine found at Brickhill. Identification uncertain.

Localities. Brickhill, Farringdon.

PSEUDODIADEMA FITTONI, Wright.

Pseudodiadema Fittoni, Wright, *Pal. Soc., British Cret. Echinoderms*, p. 90, pl. 15, f. 1.

Common at Brickhill.
Localities. Brickhill, Shanklin (Mr Meÿer's Coll.), Atherfield, (Cracker Beds, No. 4 Fitton).

PSEUDODIADEMA ROTULARE, Agassiz.

Diadema rotulare, Ag.
    „   *ornatum*, Ag.
    „   *macrostoma*, Ag. and Desor.
    „   *dubium*, Sharpe.

Localities. Brickhill, Upware, Farringdon.
N. Europe. Achem. Hils formation, Göttingen Museum.
S. Europe. Abundant in the central Cretaceous areas of France; Middle Neocomian, also Upper Neocomian. Switzerland: St Croix, Landeron, Hauterive.
Range. Lower to Upper Neocomian.

PELTASTES WRIGHTII, Desor.

(Plate VII. fig. 10.)

Salenia punctata, Forbes (non Desor).
Hyposalenia Wrightii, Desor.
Peltastes Wrightii, Wright, *Pal. Soc. Brit. Cret. Ech.*, p. 150, pl. XXX., f. 1, 2.

Localities. Upware, Brickhill (common), Farringdon, Hythe, Atherfield, Sandown.

136

Salenia hieroglyphica, n. sp.

(Plate VII. fig. 11, *a, b, c.*)

Description. *Test* circular, inflated, the summit more or less elevated and obliquely pyramidal.

Ambulacral areas narrow and prominent, widening very slightly towards the mouth; the borders ornamented with alternating rows of large, mammillated granules, between which are scattered a few smaller, simple granules.

Poriferous zones narrow and flexuous; pores unigeminal.

Interambulacral areas wide with a double series of four or five primary tubercles, increasing in size from the peristome upwards. There is a moderately wide medial granulated zone, broader than the ambulacral zone. Interambulacral plates short, only slightly broader than high; areolæ of the tubercles large, running up bare to the poriferous zone, but leaving a narrow border with large granules above and below, and a wider, granulated area on their medial margins.

Base of test convex, the peristome being inflected.

Mouth moderate, with small peristomal notches.

Apical disc rather large, elevated, with festooned margin; its surface marked with large punctures, and groovings of varying shapes—rounded, oval and triangular pits, graduating into deep furrows. The punctures are best marked near the margin, the central region being covered with deep, radiating and transverse furrows. The specific name indicates the appearance of this part of the test.

Anus prominent, directed obliquely forwards towards the axial (anterior) ambulacrum. Periproct triangular, strongly margined.

The *tubercles* are of moderate size; boss broad and gentle, with coarsely granulated summit; the mammilla small. The granules are of three orders: (i) the large bordering granules with very large mammilla, and narrow-rimmed, steep boss; (ii) smaller simple granules sparsely scattered; and (iii) miliary granules, seen abundantly amongst the larger types in the medial ambulacral and interambulacral zones.

Measurements. Diameter, 18 mm.; height, 13 mm.; diameter of apical disc, 11 mm.

POLYZOA. 137

Affinities and differences. This Urchin is remarkably similar to the *Peltastes Studeri*[1] (Cotteau) of the French Aptien rocks, agreeing with that species in general shape and size, in the grooved character of the apical disc, and also in much of the detail of the tubercles; but our Upware species is a true *Salenia*, not a *Peltastes;* and the median zone of tubercles in the interambulacral area is somewhat broader, and the tubercles closer packed than in *P. Studeri.* Again, the latter has straight ambulacral areas, and possesses a series of six tubercles.

The nature of the apical disc, especially its highly developed grooving and the prominent anal aperture serve to distinguish this from other species of *Salenia.*

Locality. Brickhill.

POLYZOA.

(See *ante*, p. 27.)

REPTOMULTISPARSA HAIMEANA, de Loriol.

Reptomultisparsa Haimeana, de Loriol, *Anim. Invert. Mont. Salève*, p. 136, pl. 17, f. 4.

Our Upware specimen (only one is known, Woodw. Mus.) is finely developed, the mass measuring three inches in length. De Loriol's type is about one inch across.

Localities. Upware.

S. Europe. Mont Salève, Middle Neocomian.

ENTALOPHORA RAMOSISSIMA, d'Orbigny.

Entalophora ramosissima, d'Orbigny, *Pal. Fr. Terr. Crét.* v. p. 785, pl. 618, figs. 1—5.

Localities. Upware, Brickhill, Farringdon.

S. Europe. Villers, Calvados (Cénomanien).

ENTALOPHORA ANGUSTA, d'Orbigny?

Entalophora gracilis, d'Orb. *Pal. Fr. Terr. Crét.* v. pl. 617, f. 1—4. *E. angusta,* p. 783.

In our species the cells are more definitely bounded and closer packed than in d'Orbigny's figures.

[1] Cotteau, *Pal. Française, Terrains Crétacés,* Ech. II. p. 111, pl. 1026.

Locality. Upware.

E. angusta d'Orb. is from Grand pré, Ardennes (Albien).

ENTALOPHORA DENDROIDEA, n. sp.

(Plate VII. fig. 12, *a, b.*)

A stout dendroid Polyzoarium, composed of many layers of superposed polypides. The branches are round in section, and they divide irregularly, sometimes dichotomously.

The cell apertures are small and but slightly projecting; they are spirally arranged, in quincunx order, and the interspaces are wide and smooth. A faint outline of the contours of the cell bodies may sometimes be seen as in *Multelia irregularis* (d'Orb.).

Affinities and differences. The many layers of cells of which the colony is composed distinguish this from all the other species of *Entalophora.* Indeed it is a character of generic value according to M. d'Orbigny's system. The large size and stoutness of the colony are also very characteristic. The peculiar mode of growth of *Multelia* which produces successive zones of different thickness at once distinguishes all the forms of that genus from our species.

Localities. Upware, Brickhill.

MELICERTITES UPWARENSIS, n. sp.

(Plate VII. fig. 13, *a, b.*)

A large branching species, forming a colony like that of the common *Entalophora ramossissima.* The branches frequently anastomose.

Cell apertures large, subtriangular, the base straight or slightly convex towards the cavity, sides simple, apex rounded. Each cell aperture is bounded by a definite ridge. The cells are rather close packed and arranged in spiral lines.

The large size and anastomosing character of the colony, the shape of the cell apertures, and their denser packing serve to distinguish this from *M. Meudonensis,* d'Orb.

SEMIMULTICAVEA (RADIOPORA) TUBERCULATA, d'Orbigny.

Semimulticavea tuberculata, d'Orb. *Pal. Fr. Ter. Crét.* v. p. 980, pl. 648, f. 1—4.

A somewhat larger and coarser type than that figured by d'Orbigny.

Localities. Upware, Farringdon.

S. Europe. Le Havre, in the Cénomanien beds.

RADIOPORA BULBOSA, d'Orbigny, var.

Radiopora bulbosa, d'Orb. *Palaeont. Francais Terr. Crét.* p. 996, pl. 650, f. 6—8.

Our specimens are much larger than (four times as large as) that figured by d'Orbigny, and the base is deeply concave, so that the colony is a hollow hemisphere instead of a pyriform bulb, as in d'Orbigny's type. Without seeing more specimens I cannot consider these differences as of more than varietal value, and therefore name our specimens *Radiopora bulbosa,* var. *magna* (Keeping).

Localities. Upware.

S. Europe. Le Mans (Sarthe), Cénomanien.

CERIOPORA (ECHINOCAVA) RAULINI, Michelin.

Ceriopora Raulini, Mich. 1841, *Icon. Zooph.* p. 2, pl. 1, f. 7.

Echinocava Raulini, d'Orb. *Pal. Fr. Terr. Crét.* v. p. 1013, pl. 788, f. 7, 8.

A beautiful, coralloid, obtusely-spinose form of Ceriopora, not uncommon at Upware. The echinoid processes seem to be composed of less durable material than the main mass, so that in some specimens they have been worn away into crater-like hollows, very similar to the calices of a coral.

Localities. Upware.

S. Europe. Grandprè, Ardennes.

CERIOPORA (REPTOMULTICAVA) MAMMILLA, Reuss.

Reptomulticava mammilla, d'Orb. *Pal. Fr. Terr. Crét.* v. p. 1041, pl. 793, f. 3, 4.

Localities. Brickhill, Farringdon.

N. Europe. Bilin, Bohemia.

S. Europe. Depts. Loire and Charente Infr. (Senonien).

CERIOPORA (REPTONODICAVA) NODOSA, n. sp.

(Plate VII. fig. 14, *a, b.*)

Ceriopora cavernosa, Catalogue Mus. Practical Geol. 1878, p. 6.

Description. Polyzoarium massive and irregular, forming large amorphous, globose, and lobose masses; or pointed processes may also be developed. It is composed of an infinite number of thin superposed layers, which rest in contact one above the other. These layers are not simple and even, but are folded up into domose, mammilated and cylindrical prominences, so that the whole surface is coarsely nodulose. This character is not unfrequently much exaggerated, especially in young specimens, so that the whole structure presents a stunted, domose appearance.

The under surface is concave and shews a delicate lamellose common plate.

Cell apertures slightly contracted, rounded or subtriangular.

Our specimens (Woodw. Mus.) from Upware attain the size of 8½ × 6 × 4 inches.

Commonly the surface of the Polypidon is eroded, so that the cell apertures appear large, simple and polygonal.

Affinities and differences. This species is most nearly allied to *Ceriopora mammillosa* (Römer), from which it differs in its larger growth, and more prominent *mammillæ;* also the cell openings in Roemer's species are much smaller than in ours: "Ohne Vergrösserung nicht Sichtbaren" (Römer, *Kr.* p. 23). *Ceriopora (Reptonodicava) cavernosa* is a much smaller structure than *C. nodosa,* with minute cells; it is well separated from our Upware fossil by the hollow cavernous spaces which are left between the various superposed layers. In *C. nodosa* the layers are perfectly in contact. In *C. spongiosa,* Röm., the surface is described as smooth and the pores are round. The general aspect of this massive polyzoon is similar to *Zonopora tuberculata,* but that species has intermediate pores between the cells, and the latter have a radiating arrangement.

Localities. Upware, Brickhill.

N. Europe. Schöppenstedt, Brunswick.

REPTOCEA LOBOSA, sp. nov.

(Plate VII. fig. 15, a. b.)

Description. Polyzoarium encrusting, but developing upwards into more or less clavate lobes, which may again divide into bi- or tri-lobed prominences. A somewhat botryoidal aspect may thus be produced.

The cell apertures are obliquely set, expanded and trumpet-mouthed; in outline sub-triangular or broadly pyriform. They stand in quincunx order.

The form of the whole cell-colony, as well as the shape of the cell-apertures, at once distinguish this species.

Locality. Brickhill.

HETEROPORA (MULTIZONOPORA) RAMOSA, Roemer.

Heteropora ramosa, Röm. 1836. *Oolit. Supplement*, pl. 17, f. 17.
Ceriopora arborea, Dunker and Koch.
Multizonopora ramosa, d'Orb. *Pal. Fr. Terr. Crét.* v. p. 927,
pl. 772, f. 1—3.

The semizonary arrangement of the cell-mouths is not always clear, but may be seen in many specimens.

Localities. Upware.

N. Europe. Schöppenstedt, Brunswick.

S. Europe. St Croix, Yonne, Haute Marne.

HETEROPORA (MULTICRESCIS) MICHELINI, d'Orbigny.

Heteropora cryptopora, Michelin (non Goldfuss).
Multicrescis Michelini, d'Orb.

Localities. Upware (?), Brickhill, Farringdon.

S. Europe. Grandpré, Ardennes (Albien, d'Orb.).

HETEROPORA COALESCENS, Reuss.

Heteropora coalescens, Reuss, in Geinitz, *Palæontographica*, 1871,
I. p. 131, pl. XXXII., f. 10—12.

A zonary arrangement of the cell apertures may be detected over portions of the branches, but for the most part the cells are scattered as in Geinitz's figure.

Localities. Upware, Brickhill.

N. Europe. Dresden (Untere Quader).

HETEROPORA (MULTICRESCIS), sp.

Allied to *Heteropora cryptopora*, Goldfuss, but the colony is more massive and less branching and the cells are much larger. It is near to *Multicrescis variabilis*, d'Orb., but that is a much smaller Cenomanian species, branching, and not spreading and massive as is the habit of the Wicken form.

Locality. Upware.

HETEROPORA (MUTICRESCIS) sp.

Grows apparently in the form of massive cylindrical or pyriform, slightly flattened branches. Only one specimen is known, which measures $4\frac{1}{2} \times 1\frac{1}{2} \times \frac{3}{4}$ in. greatest length, breadth and thickness. There is an irregular swelling at the upper third of its length, where a lateral stem was given off at an acute angle. The cells are arranged in numerous superposed layers, and on the perfect surface their openings are seen to be small, and surrounded by a number of small pores.

This is allied to *Heteropora digitata*, Mich.

Locality. Upware.

HETEROPORA (NODICRESCIS) ANNULATA, sp. nov.

(Plate VII. fig. *a*, *b*.)

Polyzoarium fixed by an expanded base, whence arise a number of stout, irregular, cylindrical, terete trunks.

These trunks are commonly united in their lower part into a stout, irregular, upright plate or amorphous mass, presenting rounded protuberances.

The free trunks are cylindrical and massive, usually curved and sometimes anastomosing; they here and there give off lateral protuberances and stout branches.

The whole surface is covered with small ridges and prominences, arranged confluently into lines, running more or less transversely, so as to produce an annular, obliquely-annular, or spiral appearance over the whole body.

Cell-mouths small, rounded or squarish, irregularly scattered, with few and variable intermediate pores of about half the size of the cell apertures.

Our specimens (Woodw. Mus.) attain a height and breadth of 3—4 inches, the branches being a half and three-quarter inches in diameter.

Locality. Upware.

Affinities and differences. *Nodicrescis tuberculata* (D'Orb. *Terr. Crét.* v., pl. 800), and *N. verrucosa*, (Römer, *Kreide*, pl. v., f. 26, p. 23), are allied to this remarkable Upware species, but neither of them present the characteristic confluence and oblique-ringed arrangement of *N. annulata.* *Ceripora papularia* is very similar in *gross* appearance, though a more slender and anasto-mosing structure, but according to d'Orbigny, it is quite a distinct genus (*Reptomulticlausa*).

HETEROPORA (REPTONODICRESCIS), sp.

An encrusting species growing upon an oyster or *Plicatula.* The cells are large and close set, and the intermediate pores are sparsely scattered between them. It may be the young and rep-tant condition of *H. annulata*, but the cell openings are rather larger.

Locality. Upware.

HETEROPORA (MULTICRESCIS) ARBUSCULA, n. sp.
(Plate VII. fig. 17 *a, b.*)

Ceriopora polymorpha, Cat. Mus. Pract. Geology, 1878, p. 7.

Description. Polyzoarium rather bush-like, attached by a broad base, from which one or many trunks may arise. Branches taper-ing, circular in section.

Over the surface are seen (i) large cell-openings, circular, and with slightly prominent peristome ; these are either very few and irregularly scattered, or grouped in patches, or arranged in irregular zones. The areas between these are occupied by (ii) the numerous small pores which are about half the diameter of the cell mouths.

In places the cells are arranged in distinct spirals, but generally they seem to be quite irregular in their distribution.

The colony usually measures about one inch in height and breadth.

Localities. Upware, Brickhill.

Affinities and differences. This polyzoon has been very generally called by collectors *Ceriopora polymorpha* of Goldfuss, but that is a true *Ceriopora* (see Goldfuss, *Pet. Deutsch*, pl. x. f. 7 *d*). Moreover ours is a much more branching structure, rarely shewing any tendency to fuse into a layer. From *Heteropora ramosa* it is distinguished by its much more branched and bushy colony and by its smaller cell openings.

Its habit of growth distinguishes it readily from *Heteropora coalescens*, Reuss, whose branches fuse together into an anastomosing mass of stems. *Sparsicavea irregularis*, D'Orbigny, bears a close resemblance to our species, but this also forms frequent anastomoses, and its intermediate pores are fewer and less conspicuous. In *S. dichotoma*, Hag., the colony is less branching and the cells are much smaller.

Localities. Upware, Brickhill.

HETEROPORA MAJOR, n. sp.

(Plate VII. fig. 18, *a, b.*)

Description. *Polyzoarium* dendroid; the branches irregular, tapering, in section circular or oval. Contiguous branches often fuse and anastomose so as to form wall-like expansions. Cell apertures numerous, simple, and small; the peristome is plane with the general surface. Intermediate pores few and small, irregularly scattered.

Affinities and differences. Distinguished from *H. (Zonopora) ramosa* (Römer), and *H. lœvigata*, d'Orb., by the small size of its cell openings, their wide separation, and the small, scattered intermediate pores. The colony is larger than commonly occurs in *H. ramosa* (about twice as big), and I have not detected the zonary arrangement of the cells such as occurs in the two above-named species. It is nearer to *H. Buskana* (de Loriol, *Foss. Mont Salève*, pl. XVIII. f. 6), but its stouter growth and the smaller size of its cells serve to distinguish it also from this species.

Locality. Upware.

PORIFERA.

Sponges.

For a general account of the Sponges see *ante*, pp. 28, 29 (*Indigenous fauna*).

The structure of most of our species is illustrated by the work of Prof. Sollas upon the 'Catagmidæ'[1].

Whatever may have been the original constitution of these sponges they must have been originally hard and rigid, for large *Serpula* and other episites often affixed themselves to them. The fibre is well seen in the coarse-formed *Pachytiloda*, shewing simple spicular fibres, having no relation with the *Lithistida*.

The name *Catagma* (Sollas) is here used by me in a more restricted sense than that originally proposed by the author, whose diagnosis seems to me to include too large and diverse a group for a single genus. Therefore I have followed Zittel in retaining the old name *Elasmostoma* for such osculated forms as *E. peziza* and *E. macropora*, and I adopt *Catagma* for the remaining Upware forms without distinct oscules, as *C. cupuliformis* and *C. porcatum*. A new name was necessary for the latter group because of the previous confusion, but the former genus *Elasmostoma* has remained a well-defined and, as I think, natural genus since its first establishment[2].

HEXACTINELLIDA.

Plococyphea pertusa, Geinitz, *das Elbthale Sachsen*, I. p. 26, pl. II., f. 5.

(Plate VIII., fig. *a*, *b*.)

The hexactinellid structure of this sponge is beautifully preserved in carbonate of lime, and in places it is well seen left in relief.

Locality. Brickhill.

N. Europe. Near Dresden (Untersten Planer).

VERTICELLITES ANASTOMOSANS, Mantell, *Medals of Creation*, p. 227.

See Sharpe, *Quart. Journ. Geol. Soc.*, X., p. 195, pl. V., f. 1.

Localities. Upware, Brickhill, Farringdon, Godalming, Atherfield.

S. Europe. Ardennes.

[1] Annals and Magazine of Natural History, Nov. 1878.
[2] Professor Sollas writes me his entire approval of the arrangement here made.

VERTICELLITES ANNULATUS, sp. nov.

(Plate VIII., fig. 2.)

Sponge bodies cylindrical, growing attached in groups. The segments of the bodies are numerous and distinctly defined, producing a strong annulation of the exterior; distal (oscular) end gently convex, flattened. The whole surface is strongly granulated. *Measurements.* Our specimens are about 20 mm. long and 10 mm. broad; height of single rings (segments) about 2 mm.

The rugose surface and strongly-ringed characters of this form distinguish it sharply from all the other species. Each sponge body looks like a series of inverted saucers, or shallow acorn-cups piled one upon the other and strung together by a central tube.

Locality. Upware.

VERTICELLITES CLAVATUS, sp. nov.

(Plate VIII., fig. 3.)

Sponges growing in groups, attached to foreign bodies. Sponge bodies simple or giving off lateral shoots; contiguous bodies often fusing together. Individuals obconic or clavate; surface smooth or irregularly annulated, shewing a delicate net-structure. Upper end gently convex with slightly projecting central osculum. Walls thin; body-cavity subdivided by numerous, delicate, convex partitions, which are connected by a central cloacal tube, provided with openings into each chamber.

Measurements. Length, $1\frac{3}{4}$ inches; thickness near summit, 15 mm.

The shape of the sponge body distinguishes this from all the other species. The walls too are thinner and the septa more delicate than in the common *V. anastomosans*, these characters being perfectly constant in all our specimens.

Localities. Upware, Brickhill, Farringdon.

CATAGMA CUPULIFORMIS, From., sp.

Chenendopora fungiformis ? Sharpe.
Cupulochonia cupuliformis, E. de Fromentell. *Introd. à l'étude des éponges Fossiles,* pl. III., f. 5.

The tissue is rather more delicate than in the other Upware species. There is no well-marked epidermal layer.

Localities. Upware, Brickhill ?

N. Europe. Plauen, Gamughagel, Dresden.

S. Europe. Mont Salève, Germigney.

CATAGMA PORCATUM, Sharpe, sp.

Manon porcatum, Sharpe, *Quart. Journ. Geol. Soc.,* x., p. 196, pl. v., f. 2.

This species forms less perfect cups than the *Elasmostoma acutimargo,* large expansions and folded plates being most frequent. Still the cup is its proper plan of growth, and very perfect cups do occur.

The outer surface is covered with irregular, wart-like, branching or crumpled prominences, which may assume an imperfect radiating arrangement towards the margin; its surface shews the simple bare vermiculate structure. On the inner surface the epidermal layer is dense and compact, pierced by an infinite number of pore-like current apertures (oscules). The epidermal layer is often destroyed, leaving the vermiculate structure exposed. There is no definite arrangement of the oscules.

Localities. Upware, Brickhill, Farringdon.

ELASMOSTOMA ACUTIMARGO, Römer.

Tragos acutimargo, Römer, Vol. ii. 10, pl. xvii., f. 26.

Manon macropora, Sharpe, *Quart. Journ. Geol. Soc.,* x., p. 195, pl. v., figs. 3, 4.

Very fine specimens occur of all sizes up to a cup seven inches in diameter. Small specimens may be quite irregular little plates, or fans, often with the oscules external; but all our large specimens are cups, though often very irregular. The outer (pore) surface is smooth and simple, or with faintly marked, gentle, radiating crumplings and concentric growth-lines. On the inside of the cup are seen the numerous large oscules arranged in distinct rings. A cup measuring 3½ inches diameter and 2 inches high has eleven such rings. The oscules vary in size from 1—4 mm. in diameter; and they may be simple openings, in the substance of the sponge,

or projecting as prominent craters or tubes (*var. tubifera*, Kpng.).
Two specimens in the Woodwardian Museum have a well-marked,
expanded, flattened, and everted margin, and these specimens
further agree in being of stouter build than is usual in the ordinary
forms thus agreeing with the type of *E. marginata*.

The vermiculate structure of the sponge is well seen over the
outer surface, but on the oscular face the tissue is denser, forming
an '*epitheca*' which is however less developed than in the Farringdon sponges.

Localities. Upware, Brickhill, Farringdon.

N. Europe. Schöppenstedt (Hils conglom.), Dresden
(Planer)?

S. Europe. Ardennes, Arzier (Vaud).

ELASMOSTOMA PEZISA, Goldfuss.

Manon pezisa, Goldfuss, *Petr. Germ.*, I., pl. I., figs. 7, 8 (not the
other figures).

Elasmostoma consobrinum, d'Orb., Geinitz, and others.
Localities. Brickhill, Upware, Farringdon.

CORYNELLA NODOSA, sp. nov.

(Plate VIII., fig. 4.)

Description. Sponges more or less pyramidal, rarely cylindrical; attached by a broad base; sessile, or mounted upon a short
neck.

Walls very thick, composed of an uniform, coarse vermiculate
tissue, leaving only a narrow cylindrical central cloaca.

The system of canals opening into the central cloaca is but
imperfectly developed.

Osculum terminal, central.

A simple smooth area forms the terminal cone around the
osculum, but below this the surface is raised into prominent, coarse
knobs and foldings.

Locality. Brickhill.

CORYNELLA, sp.

(Plate VIII., fig. 5.)

This is likely to be the young of *C. nodosa,* Kpng.
Locality. Brickhill.

PERONELLA FURCATA, Goldf., sp.

Scyphea furcata, Goldfuss, *Petr. Germ.* I., pl. II., f. 6, p. 5.
Polyendostoma furcata, Römer.
Cnemidium astrophorum, Mantell, *Medals of Creation,* p. 227.

Localities. Brickhill, Farringdon.
N. Europe. Essen.
S. Europe. Ardennes, Mont Salève.

PACHYTILODA, Zittel.

(Plate VIII., fig. 6.)

A remarkably coarse sponge structure, with open meshes 4 mm. across, and even some of them as much as three-quarter in. long.
Diameter of fibres variable (1—3 mm.).

The fibres usually form an irregular network, but sometimes expand into lamellæ; numerous fibres may thus unite to form plates running in any direction.

The most perfect specimen, from Brickhill, is a hemispherical mass, about two inches in diameter. The basal fibres are fused into an irregular lamina, fixed to fragments of rock for attachment; patches of lateral surface laminæ were also locally developed, and irregular plates are included in its substance.

There is no trace of canals, oscules, or cloaca, the water circulation having apparently no definite courses.

Constitution of the fibre. The Upware specimens, which are of chalky consistency, exhibit the coarse fibres as made up of delicate curved filaments (spicules) lying parallel with each other and arching out from the centre towards the circumference, thus presenting the appearance of a branch of small-celled *Chœtetes.* This structure is visible to the naked eye, and is best seen with a pocket lens.

Localities. Upware, Brickhill.

N. Europe. Precisely similar masses of coarse sponge tissue occur at Essen (" *Cupulospongia infundibuliformis* ") ; and at Plauen near Dresden.

PLANTÆ.

Small ferruginised fruits, with overlapping scale-like bracts. Affinities doubtful.

See Plate VIII., fig. 7.

Locality. Upware.

DERIVED FOSSILS.

A. THE DERIVED FOSSILS OF NEOCOMIAN AGE, FOUND IN THE DARK IRONY GRIT. (See *ante*, p. 32.)

Cerithium (Turritella) granulata, Forbes ; *non* Sow.

Localities. Upware ; Ingoldsthorpe, Norfolk (Geological Society's Collection).

Trochus, sp.

(Ornamented species.)

Locality. Upware.

PERNA RICORDIANA, d'Orbigny.

(Hinge portions of.)

Locality. Upware.

This species occurs in the Atherfield beds, Isle of Wight, and in the Shanklin Sands ; also in the Middle Neocomian of Tealby, in Lincolnshire, and of Villers le Lac, St Croix, and in the Aptien of the Perte du Rhone.

PERNA MULLETI, Deshayes.

Casts of the ligament area, shewing the alternating larger and smaller fossæ as figured by MM. Leymerie and Forbes.

Localities. Upware, in the black grit nodules ; base of the Atherfield series ; Maidstone ; Surrey ; Tealby, middle series ; Speeton, upper series.

N. Europe. Schöppenstedt, Hils conglomerat.

S. Europe. Switzerland, Valengien to Urgonien; Paris area, Neocomian.

THETIS MINOR, Sby.

Thetis minor, Sby. Characteristic specimens occur not unfrequently in the Upware grit boulders, associated with *Trigonia ornata, Litorina,* sp. *Pecten orbicularis,* and a small sub-triangular bivalve.

It is more common at Potton.

Localities. Shanklin; Wiltshire (Morris); Speeton (Upper Neocomian).

TRIGONIA ORNATA, d'Orb. (Lycett).

Trigonia ornata, d'Orbigny—Lycett. *Brit. Foss. Trigonia, Pal. Soc.,* p. 139, pl. XXIV., figs. 6, 7.

Localities. Hythe; Atherfield, in the Perna bed.

S. Europe. Perte du Rhone, Aptien; Paris area.

TRIGONIA SPINOSA, Park. ?

Trigonia spinosa, Park.? Lycett, *Brit. Foss. Trigonia, Pal. Soc.* p. 136, pl. XXIII., f. 10; pl. XXIV., figs. 8, 9.

Localities. Brickhill; Blackdown beds; Isle of Wight (Upper Greensand).

CUCULLÆA VAGANS[1], sp. nov.

(Plate VIII., fig. 8, *a, b.*)

Shell globose, nearly equilateral, ornamented with strong, radiating ribs. Umbones prominent, rounded, slightly anterior. Hinge line straight; teeth few, oblique in the cardinal region, longitudinal at the sides; the lateral ones are striated across. Area triangular, moderate, with angular cartilage grooves.

Ribs about 25 in number, strong and equal over the back, faint in the cardino-lateral region; they are well defined, angular in cross section at the sides, but more rounded over the back. Secondary striæ usually run along upon the ribs. Interspaces broader than the ribs, widest at the sides, often with faint radiating

[1] *Vagari,* to wander. It occurs in scattered blocks, being unknown *in situ.*

striæ. A set of well-defined regular concentric striæ crosses the ribs and interspaces, producing a distinct scalariform structure. The cast is smooth or faintly radiated, with a posterior groove left by the muscle-shelf.

Measurements. Cardino-pallial diameter, 19 mm.; antero-posterior diameter, 21 mm.; thickness (through both valves), 14 mm.

Affinities and differences. This shell is so very nearly equilateral that it might be referred to the genus *Pectunculus*, but it has the projecting shelly muscular shelf for the insertion of the posterior adductor muscle as in *Cucullæa*. Its globose, nearly equilateral form, sharp ribs with few striations, and the regular cross-barring of the interspaces well distinguish this species.

From *C. rotundata*, Rœmer, it is distinguished by its larger size, better-defined ribs, and larger more rounded umbo.

I have little doubt that the original home of this shell was in Neocomian waters, though now it is only known as a derived fossil in newer Neocomian deposits.

Localities. In the Black Grit rock, derived blocks, Upware; also in the Herrimere Rock.

CUCULLÆA DONNINGTONENSIS, sp. nov.

(Plate VIII., fig. 9, *a, b.*)

Shell globose-quadrate, inequilateral, radiately ribbed. Umbones prominent, rounded, oblique. Anterior side rounded; posterior obliquely truncated. Hinge line short; area narrow, marked with angular cartilage grooves. Ribs about 30, equal but becoming faint towards the posterior side; they are angular in section, moderately prominent, and covered with strong secondary striæ, which also occupy the interspaces. These are crossed by close-set growth striæ.

Measurements (type from Upware). Cardino-pallial diameter, 21 mm.; antero-posterior diameter, 25 mm.; hinge-line, 12 mm.; thickness (through both valves), about 20 mm.

A ribbed species more oblique than the last, and with the ribs more triangular in section and covered, together with the interspaces, with numerous and well-marked striæ, which are crossed by close-set lines of growth; also the umbo is less prominent.

In these characters it approaches the fine species so abundant in the middle series at Tealby, but its more globose and inequilateral form and less square outline at once distinguish it from that species.

Localities. I only know of one specimen (W. M.) from the black grit nodules of Upware, and one well preserved shell from the Lower Cretaceous sands of Doddington in Lincolnshire.

B. THE DERIVED FOSSILS OF NEOCOMIAN AGE.

(See *ante*, p. 31.)

ANCYLOCERAS ?

Scattered joints, nearly one inch in diameter, are occasionally found at Upware. In the Potton beds much larger specimens occur, which have been referred, with good reason, to *Ancyloceras gigas*. Similar large fragments, likewise in a phosphatised condition, may be picked up on the shore in Heligoland, where they are washed out of the Speetonian clays.

ANCYLOCERAS, sp. 2.

Two phosphatised fragmentary specimens of this species have occurred in the Upware deposit. They are about 1½ inches long and ½ inch in greatest diameter. Section circular or elliptical. The ribs are simple, bifurcate or trifurcate, and somewhat irregular in size ; some of the ribs are simple throughout. All the ribs pass regularly over the back, but are only represented by delicate lines on the concave side.

Locality. Upware.

AMMONITES DESHAYESII, Leym.

This is a not uncommon fossil amongst the phosphatic nodules of Upware.

It is undoubtedly a derived fossil, having frequently been broken up and rolled into pebbles on the Neocomian shore at Upware. All the specimens are in a phosphatised condition—a gritty phosphate—and its mode of occurrence and general condition are very similar to the *Ammonites biplex.*

Localities. Upware, Atherfield, Speeton (Upper Neocomian).

N. Europe. Brunswick, Upper Neocomian (Judd).

S. Europe. Paris area, Aptien.

HAMITES ?

Fragment of a small round-sectioned Hamite, with prominent rounded ribs, which are mostly simple—only one of them is seen to bifurcate. They are separated from one another by more than their own breadth.

Locality. Upware (Coll. Mr J. F. Walker, of York).

? GERVILLIA SOLENOIDES, Defrance.

Locality. Upware (Coll. Mr J. F. Walker, M.A.).

TEREBRATULA OVOIDES, Sowerby.

Variety *rex*, Ray Lankester.

(Plate VIII., fig. 10, *a*, *b*.)

Terebratula ovoides, Sby., *Min. Con.*, t. 100, p. 227.

Terebratula rex, E. Ray Lankester, *Geol. Mag.*, 1870, p. 410.

A species to which much doubt and considerable interest attaches, on account of its wide distribution in boulders, while it is unknown *in situ.* It is a large species, with a very characteristic medial ridge along the large valve, and with the dorsal valve frequently flattened in the specimens from Upware, S. Willingham, and Herrimere.

The lateral beak ridges may be sharp and the lines of growth may become developed into imbricating squamæ.

Sowerby's original specimens were from a sandstone boulder, " a sandstone containing green sand " found in Suffolk (*T. ovoides* and *T. lata*, Sow., *Min. Con.*, I. p. 227), and I have found it in a hard green-grey sandstone boulder at South Willingham in Lincolnshire, also at West Dereham, near Downham Market.

But its most interesting occurrence is at Herrimere, N. of Upware on the Cam, for here it occurs associated with other fossils whose age can be determined, namely, Lower Cretaceous. I have already (see *ante*, p. 35), given my reasons for believing that this occurrence at Herrimere is not *in situ*, as has been stated elsewhere. (*Geol. Mag.*, 1870), nor of Jurassic age, but that it is really a boulder of Neocomian sandstone.

The matrix in all the known localities, except Upware, appears to be similar, namely, a hard green-grey sandstone most like the hard consolidated masses in the Lowest Cretaceous Sands of Lincolnshire.

At Upware it occurs as internal casts in a compact earthy phosphate of lime, often much rolled and penetrated by boring bivalves.

Localities. Upware, Brickhill, Potton.

In the gravels and drift of Suffolk and Norfolk.

Also in boulders in Lincolnshire (S. Willingham), and West Dereham; Downham Market; Ely; also Feltwell in Norfolk.

C. DERIVED FOSSILS BELONGING ORIGINALLY TO THE JURASSIC AGE.

(See *ante*, pp. 41—46.)

(They occur as a rule in a phosphatised and much rolled condition.)

Belemnites explanatus, Phillips. *Monog. Belemnites, Pal. Society,* t. XXXVI., figs. 94—96, p. 128.

In the alveolar cavities groups of *Serpula tricarinata* (a Jurassic fossil) are found adhering.

The species is abundant in the Ampthill Clay, and in the Kimmeridge Clay of Hartwell, Wheatley, &c.

Locality. Upware.

AMMONITES GIGANTEUS, Sby. ?

Fragments of very large Ammonites which probably belong to this species are found in the Brickhill area.

AMMONITES BIPLEX, Sowerby.

(Plate VIII., fig. 11.)

Typical forms of this species abound in all the Neocomian phosphatic deposits of Cambridgeshire and Bedfordshire, and they are also conspicuous in the similar beds in Lincolnshire. In some of the pits, especially near Brickhill, they actually form a large proportion of the heaps of phosphatic nodules. They are invariably phosphatised, and the outer whorls have usually been worn by attrition, so that fair-looking specimens can only be obtained by breaking away the outer whorls. Many of the shells have been utterly broken up and rolled into pebbles.

A broad-whorled, many-ribbed variety of this species also occurs in excellent preservation at Brickhill, Upware, and Potton.

From their occurrence with *Astarte cuneata, Trigonia incurva*, and other fossils of Portlandian and Purbeck age, and from the nature of the phosphate in other specimens, I conclude that most of these Ammonites were derived from old Portlandian and Purbeck rocks, but others were I believe gathered from the Kimmeridge Clay, whilst some few are of Oxfordian types.

AMMONITES BIPLEX, d'Orb., *non* Sby.

A well-marked Oxford Clay form with narrow whorls and depressed bands, or *varices* running at intervals around them. This species, as also *Ammonites cordatus*, var. *Lamberti*, and *Ammonites cordatus*, var. *Mariæ*—all of them from the Oxford Clay, invariably occurs mineralized in oxide and sulphide of iron.

The remains of *Gasteropod* shells are rather numerous, but they are much worn, and but few of them can be specifically identified. We recognise the following :

Chemnitzia Heddingtonensis, Sby.	*Nerinæa.*
Alaria. Several species.	*Pleurotomaria reticulata*, Sby.
Cerithium muricatum, Sby. (from the Coral Rag).	*Natica.* Several species.

Other Jurassic species are :

Lamellibranchiata :

Ostrea gregaria, Sby.	*Cyrena rugosa*, Sow.
Pectunculus, sp.	*Myoconcha Portlandica*, Blake.
Cardium striatulum, Sby.?	*Myacites cuneata*, Phill.
Lucina Portlandica, Sby.	„ *parallellus*, Phill.
Trigonia gibbosa, Sby.	*Pholadomya tumida*, Ag.
„ *incurva*, Sby.	*Pholadidea* (Kimmeridge clay species).
Astarte cuneata, Sby.	
„ *Hartwelliensis*, Sby.	

Echinodermata :

Apiocrinus ? column of.

TABLES OF
UPWARE AND BRICKHILL FOSSILS.

	S. Europe	Volcras	Landeron	Mont Salève	Perte du Rhône	St Croix	Mediterranean and Rhône Basins	Switzerland	Paris Basin	Ardennes	N. Europe	Brunswick	Hunstanton Red Chalk	Atherfield	Shanklin	Blackdown	Folkestone Gault	Tealby	Speeton	Blythe Beds	Folkestone Beds	Godalming	Farringdon	Potton	Brickhill	Upware
VERTEBRATA																										
Fishes																										
Sphaerodus Neocomiensis (Ag.)			+					+	+																+	+
Otodus or Oxyrhina (teeth and vertebræ)			+					+		+															+	+
Pycnodus Couloni (Agassiz)										+															+	+
Gyrodus																									+	+
Strophodus										+															+	+
Hybodus with Sphenonchus																									+	+
Ischyodus Townsendi																									+	+
Acrodus																									+	+
Asteracanthus																									+	+
Reptiles																										
Iguanodon																									+	+
Plesiosaurus																									+	+
Pliosaurus?																									+	+
Ichthyosaurus	+			+ +				+ + +	+		+ + +	+ + +						+	+ + +			+			+	+
Crocodilian teeth and fragments of bones	+				+			+ + +	+		+ + +	+ + +						+	+ + +	+		+	+		+	+
Dakosaurus	+ + +																									+
INVERTEBRATA																										
Cephalopoda																										
Belemnites pistilliformis (Bl.)					+				+									+	+	+		+	+		+	+
" subfusiformis (d'Orb.)																									+	
" subquadratus (Röm.)																										
" Upwarensis (Keeping)																										+

+	+	:	:		+	+	:	:	:	:	:	:	:	:	:	:	:	:	:	:	:		+	:	:	+	+	:	+	+	:	+	+
:	:	:	:		:	:	:	:	:	:	:	:	:	:	:	:	:	:	:	:	:		:	:	:	:	:	:	:	:	:	:	:
:	:	:	:		:	:	:	:	:	:	:	:	:	:	:	:	:	:	:	:	:		+	:	:	+	:	:	:	:	:	:	:
:	:	:	:		:	:	:	:	:	:	:	:	:	:	:	:	:	:	:	:	:		+	:	:	+	:	:	:	:	:	+	:
+	:	:	:		:	:	:	:	:	:	:	:	:	+	:	:	:	:	:	:	:		+	:	:	:	:	:	:	:	:	:	+
:	:	:	:		:	:	:	:	:	:	:	:	:	:	:	:	:	:	:	:	:		:	:	:	:	:	+	+	:	:	:	:
:	:	:	:		:	:	:	:	:	:	:	:	:	:	:	:	:	:	:	:	:		:	:	:	:	:	:	:	:	:	:	:
+	:	:	:		:	:	:	:	:	:	:	:	:	:	:	:	:	:	:	:	:		+	:	:	+	:	:	+	+	:	+	:
:	+	:	:		:	:	:	:	:	:	:	:	:	:	:	:	:	:	:	:	:		+	:	:	+	+	:	:	:	:	+	+
:	:	:	:		:	:	:	:	:	:	:	:	?	?	:	:	:	:	:	:	:		:	:	:	+	:	:	:	:	:	+	:
+	+	:	:		:	:	:	:	:	:	:	:	+?	+?	:	+	:	:	:	:	:		+	:	:	+	+	+	+	+	:	+	:
:	+	:	:		:	:	:	:	:	:	:	:	+?	:	+	:	:	:	:	:	:		+	:	:	+	:	+	:	:	:	+	:
:	:	:	:		:	:	:	:	:	:	:	:	:	:	:	:	:	:	:	:	:		:	:	:	:	:	:	:	:	:	:	:
:	+	:	:		:	+	:	:	:	:	:	:	:	:	:	:	:	:	:	:	:		:	:	:	+	+	:	:	:	:	:	+
:	:	:	:		:	:	:	:	:	:	:	:	:	:	:	:	:	:	:	:	:		:	:	:	:	:	:	:	:	:	:	+
:	:	:	:		:	:	:	:	:	:	:	:	:	:	:	:	:	:	:	:	:		:	:	:	:	:	:	:	:	:	:	:
:	:	:	:		:	:	:	:	:	:	:	:	:	:	:	:	:	:	:	:	:		:	:	:	:	:	:	:	:	:	:	:
:	:	:	:		+	:	:	:	:	:	:	:	:	:	:	:	:	:	:	:	:		+	:	:	+	:	:	:	:	:	:	:
:	+	+	:		:	:	:	:	:	:	:	:	:	:	:	:	:	:	:	:	:		+	:	:	+	:	:	:	:	:	:	:
+	:	:	+		:	:	:	:	:	:	:	:	:	:	:	+	:	:	:	:	:		:	:	:	:	:	:	:	:	+	:	+
:	:	:	:		:	:	:	:	:	:	:	:	:	:	:	:	:	:	:	:	:		:	:	:	:	:	:	:	:	+	:	:
:	:	:	+?		+	:	:	:	:	:	:	:	:	:	:	:	:	:	:	:	:		+	+	:	:	:	:	:	:	:	:	:
:	:	:	:		:	:	:	:	:	:	:	:	:	:	:	:	+	:	:	:	:		+	+	:	+	?	:	:	:	:	:	:
:	:	:	:		:	:	:	:	:	:	:	:	:	:	:	+	:	:	:	:	:		+	+	+	:	:	:	:	+	:	:	:
:	:	:	:		:	:	:	:	:	:	:	:	:	:	:	:	:	:	:	:	:		+	+	:	+	+	:	+	+	:	:	:
+	+	+	+		+	+	+	+	+	+	+	+	:	+	+	+	+	+	+	+	+		+	+	+	+	:	+	:	+	+	+	+

Ammonites Cornuelianus (d'Orb.)
" Deshayesii (Leym.)
" sp.
Ancyloceras Hillsii (Sby.)

Gasteropoda
Cerithium Marollinum (d'Orb.)
Cerithium Neocomiense (Forbes)
Tessarolax Gardneri (Keeping)
Tridactylus Walkeri (Gardner)
Scalaria Keepingi (Gardner)
Nerinæa sp.
Nerinæa tumida (Keeping)
Trochus, sp. nov.
Pleurotomaria ferruginea (Keeping)
" Renevieri (Keeping)
" gigantea (Sby.)
Turbo ikeedi (Keeping)
Litorina varicosa (Keeping)
" Upwarensis (Keeping)
" Cantabrigensis (Keeping)
" sp.
Patella, sp.

Lamellibranchia
Exogyra Couloni (d'Orb.)
" Tombeckiana (d'Orb.)
Gryphæa dilatata? (Sby.)
Ostrea frons (Park.) var. macroptera, Sby.
" var. carinata (Sby.)
Ostrea Walkeri (Keeping)
Gryphæa vesiculosa (Sby.)
Plicatula Carteroni (d'Orb.)
" equicostata (Keeping)
Neithea atava (Röm.)
" Morrisii (P. & Ren.)

INVERTEBRATA	S. Europe	Voirons	Landeron	Mont Salève	Perte du Rhône	St Croix	Mediterranean and Rhone Basins	Switzerland	Paris Basin	Ardennes	N. Europe	Brunswick	Hunstanton Red (Tealk)	Atherfield	Shanklin	Blackdown	Folkestone Gault	Tealby	Speeton	Hythe Beds	Folkestone Beds	Godalming	Farringdon	Potton	Brickhill	Upware
Lamellibranchiata																										
Neithea ornithopus (Keeping)											+	+										~	+			+
Pecten elongatus (Lam.)	+		+		+			+	+	+	+			+	+				+			+			+	
„ Raulinianus (d'Orb.)	+			+	+			+	+	+				+							+	+	+	+		+
„ Dutemplei (d'Orb.)	+		+		+			+	+	+				+				+	+	+		+	+	+		+
„ orbicularis (Sby.) var. magnus (Keeping)			+		?			+	+				+	+						+						
Hinnites Leymerii (Desh.)	+		+		+	+		+	+		+	+		+					+						+	+
Avicula Cornueliana (d'Orb.)	+		+			+		+	+		+	+		+					+				+		+	+
? Avicula sp.																										+
Pinna Robinaldina (d'Orb.)	+		+	+	+	+		+	+	+	+	+		+				+	+			+	+		+	+
Lima Farringdonensis (Sharpe)	+		+		+	+		+		+	+	+		+	+			+	+				+		+	+
„ Tombeckiana (d'Orb.)	+				+	+		+	+		+	+						+	+				+		+	+
„ longa (Röm.)	+		+	+	+	+		+	+	+	+	+		+					+			+	+		+	+
Spondylus Römeri (Desh.)	+			+				+			+	+													+	+
Trigonia Upwarensis (Lyc.)																										+
Nucula subtrigona (K. & D.)	+		+			+		+	+	+												+				+
Arca Marullensis	+		+		+	+		+	+						+			+					+	+		+
„ Carteroni (d'Orb.)	+					+		+									+						+			+
Cucullaea subnana (P. & R.)	+															+										+
„ sp.																										+
Pectunculus sublaevis (Sby.)	+		+			+		+	+																	+
„ Marullensis (Leym.)																										+
„ obliquus (Keeping)																										+
Modiola pedernalis, Röm.?	+				+	+		+	+																	+
„ obesa (Keeping)									+						+											+

	+	+				+	+				+			+	+						+		+					+	+	~	+	
																														+		
	+	+					+														+								+		:? +	
							+					+			+	+						+									
		+					+							+							+		+								
							+																								
	+	+	+				+			+											+		+					+			
	+	+								+					+					+		+								+		
																												+		+	+	
				+		+		+		+		+			+						+		+					+	+	+ ~	
						+		+		+		+			+						+		:? +					+	+ ~ ~	
		+											+	+								+									
	+							+						+							+									+ +	
												+																			
		+					+																						+		
												+																	+		
		+					+		+										+											
	+	+										+																			
		+															+ +	+ +	+ +		+		+		+ +	+
						+										+	+ +										+	+ +
		+				+ +		+ +																	:? +	
														+ +	+ +	+ +	+ +		+ +	+ +	+ +
+ + + + + + + + + + + + + + + + + + +				+ +			+ +		+ +

Modiola sp.
Cardium Cottaldinum
    " subhillanum (Leym.)
Cardita rotundata, P. & R.?
Cypricardia striata (Gein.) sp.
    " arcadiformis (Keeping)
    " squamosa (Keeping)
Opis Neocomiensis (d'Orb.)
Astarte n. sp.
    " subdentata (Röm.)
    " sp.
Cyprina Sedgwickii (Walker)
    " angulata var. rostrata (Sby.)
    " obtusa (Keeping)
Venus Vectensis (Forbes)
Panopea plicata (Sby.)
    " gurgitis (d'Orb.)
? Thracia or Tellina
Lithodomus
Pholas (Fistulana) constricta (Phill.)

Brachiopoda
Terebratella Menardi (Lam.)
    " trifida (Meyer)
    " oblonga (Sby.)
    " Fittoni (Meyer)
    " var.
    " Davidsoni (Walker)
    " Keepingii (Walker)
Kingena rhomboidalis (Keeping)
Terebratulina striata var. elongata (Dav.)
Terebratula capillata (d'Arch.)
    " sella var. Upwarensis (Walk.)
    " " Tornasensis (d'Arch.)
    " prælonga (Sby.)
    " microtrema (Walker)

INVERTEBRATA.

Brachiopoda

Species columns:

- C1 — Terebratula Lankesteri (Walker)
- C2 — " depress. (Lam.) var. uniplicata (Walker)
- C3 — " var. cyrta (Walker)
- C4 — " var. Cantabrigensis (Walker)
- C5 — " Robertoni (d'Arch.)
- C6 — " Moutoniana (d'Arch.)
- C7 — " var. Brickhillensis (Kpng.)
- C8 — " extensa (Meÿer)
- C9 — " Seeleyi (Walker)
- C10 — " Meyeri (Walker)
- C11 — " Dallasii (Walker)
- C12 — Waldheimia celtica (Morris)
- C13 — " ? pseudojurensis (Leym.)
- C14 — " faba, (d'Orb.)
- C15 — " Juddi (Walker)
- C16 — " tamarindus (Sby.) var.
- C17 — " magna (Walker)
- C18 — " Wanklyni (Walker)
- C19 — " Woodwardi (Walker)
- C20 — " Bonneyi (Kpng.)
- C21 — Rhynchonella antidichotoma (Buv.)
- C22 — " antissima (Sby.)
- C23 — " Upwarensis (Dev.)

Locality	C1	C2	C3	C4	C5	C6	C7	C8	C9	C10	C11	C12	C13	C14	C15	C16	C17	C18	C19	C20	C21	C22	C23
S. Europe		+				+							+	+		+							
Valorons																							
Landeron					+													+					
Mont Salève												+											
Perte du Rhone																							
St Croix		+																					
Mediterranean and Rhone Basins					+																		
Switzerland					+							+	+		+								
Paris Basin												+	+		+								
Ardennes		+			+						+			+						+			
N. Europe		+			+		+				+		+	+	+								
Brunswick					+		+				+		+	+	+				?				
Hunstanton Red Chalk																							
Atherfield																							
Shanklin		+							+				+	+									
Blackdown																							
Folkestone Gault																							
Tealby		+											+	+									
Speeton		+								+		+											
Hythe Beds					+									+									
Folkestone Beds												+											
Godalming		+			+		+		?	+			+		+		+	+	+			?	+
Farringdon		+			+	+					+			+	+	+	+	+			+	+	
Potton		+	+	+							+				+				+				+
Brickhill	+	+	+	+		+	+	+	+		+	+	+	?	+	+	+		+	+	+	+	+
Upware	+	+	+	+		+		+	+	+	+	?	+		+	+	+	?	+	+	+	+	+

1	2	3	4	5	6	7	8	9	10	11	12	13	14	15	16	17	18	19	20	21	22	23	24	25	26	27	28	29	30	31	Species
		+	+	+			+		+	+	+			+	+										+	+	+				Rhynchonella Cantabrigensis (Dav.)
																															,, depressa (Sby.)
				+					+				+																		*Polyzoa*
		+									+																+				Reptomultisparsa Haimeana (de Loriol)
											+												+								Entalophora ramosissima (d'Orb.)
										+																					,, angusta (d'Orb.)?
																															,, dendroidea (Keeping)
																															Melicertites Upwarensis (Keeping)
				+		+			+											+		+									Semimulticava (Radiopora) tuberculata (d'Orb.)
+					+			+				+									+	+									Radiopora bulbosa (d'Orb.) var.
					+	+	+			+	+			+			+	+	+												Ceriopora (Echinocava) Raulini (Mich.)
						+	+			+				+			+	+	+												,, (Reptomulticava) mamilla (Reuss)
																															,, (Reptomulticava) nodosa (Keeping)
																															Reptocea lobosa (Keeping)
+	+																														Ceriopora (Reptomulticava) (Keeping)
																						+		+								Heteropora ramosa (Röm.)
																				+				+								,, Michelini (Multicrescis) (d'Orb.)
																			+				+	+								,, (Reptomodicrescis) sp.
																															,, coalescens (Reuss) (Multicrescis) sp.
+	+																							+		+					,, (Nodicrescis) annulata (Keeping)
	+		+			+			+		+					+															,, major (Keeping)
																															,, arbusonla (Keeping)
+	+		+	+				+	+	+		+		+			+						+	+	+					*Annelida*
+	+	+	+	+	+	+		+				+	+	?			+	+	+	+		+	+	+	+	+	+	+			Serpula lophioda (Goldf.)
																															,, rustica (Sby.)
																																,, ampullacea (Sby.)
																																,, articulata (Sby.)
																																,, antiquata (Sby.)
																																,, plexus (Sby.)

INVERTEBRATA.	Upware	Brickhill	Potton	Farringdon	Godalming	Folkestone Beds	Hythe Beds	Speeton	Tealby	Folkestone Gault	Blackdown	Shanklin	Atherfield	Hunstanton Red Chalk	Brunswick	N. Europe	Ardennes	Paris Basin	Switzerland	Mediterranean and Rhone Basins	St Croix	Perte du Rhône	Mont Salève	Landeron	Volrons	R. Europe
Annelida																										
Serpula gordialis (Goldf.)	+	+		+				+	+						+	+		+				+				+
Vermicularia polygonalis (Sby.)	+						+																			
" Phillipsii (Römer)	+	+		+				+	+						+	+		+								+
Echinodermata																										
Pseudodiadema Fittoni (Wright)		+										+	+					+	+		+					+
" rotulare (Ag.)	+	+		+														+	+				+	+		
Peltastes Wrightii (Desor.)	+	+		+			+						+													
Salenia hieroglyphica (Keeping)		+																								
Cidaris sp.	+	+		+													+?									
" Farringdonensis, Wr.?				+																						
" (thorny spine of)	+																									
Sponges																										
Verticellites anastomosans (Mantell)	+	+		+	+								+				+									
" clavatus (Keeping)	+	+																								
" annulatus (Keeping)		+		+																						
Peronella furcata (Goldf.)	+	+		+													+									
Elasmostoma pezita (Goldf.)	+	~		+																						
" scutimargo (Röm.)	+	+		+											+	+							+			+
Catagma porosium (Sharpe)	+	+		+																						
" cupuliformis (From.)		+		+																						
Corynella nodosa (Keeping)	+	+																								
Pachytiloda		+														+										
Plocoscyphea pertusa (Geinitz)		+																								

INDEX.

Thracia, 126
Trigonia, 113
Trochus, 95
Turbo, 97

Upware, 3, 25

Vectian, 59
Venus, 125
Vermicularia, 133

Vertebrates, 16, 40, 76
Verticellites, 145

Waldheimia, 22, 129
Walker, Mr J. F., 3, 17, 48, 58
Warminster, 51
Wealden, 40, 67

Zittel, Prof., 28

PRINTED BY C. J. CLAY. M.A. & SON, AT THE UNIVERSITY PRESS, CAMBRIDGE.

PLATE I.

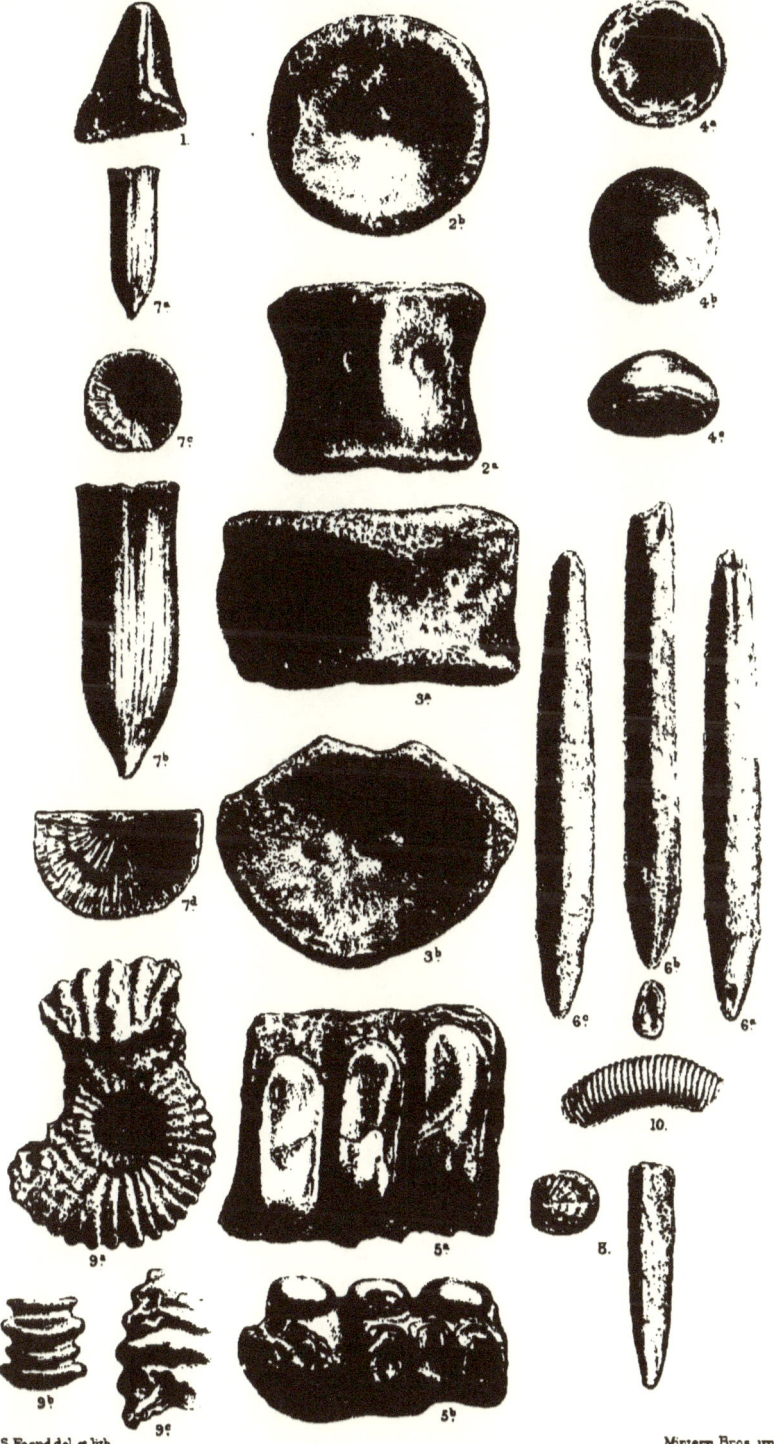

A. S. Foord del et lith. Mintern Bros imp

PLATE II.

Ancyloceras Hillsii, Sby., sp. Upware. ⅔ natural size. *b*, outline of aperture; *c*, the central part of the outer whorl; *d*, earliest part of outer whorl.

Pl. II.

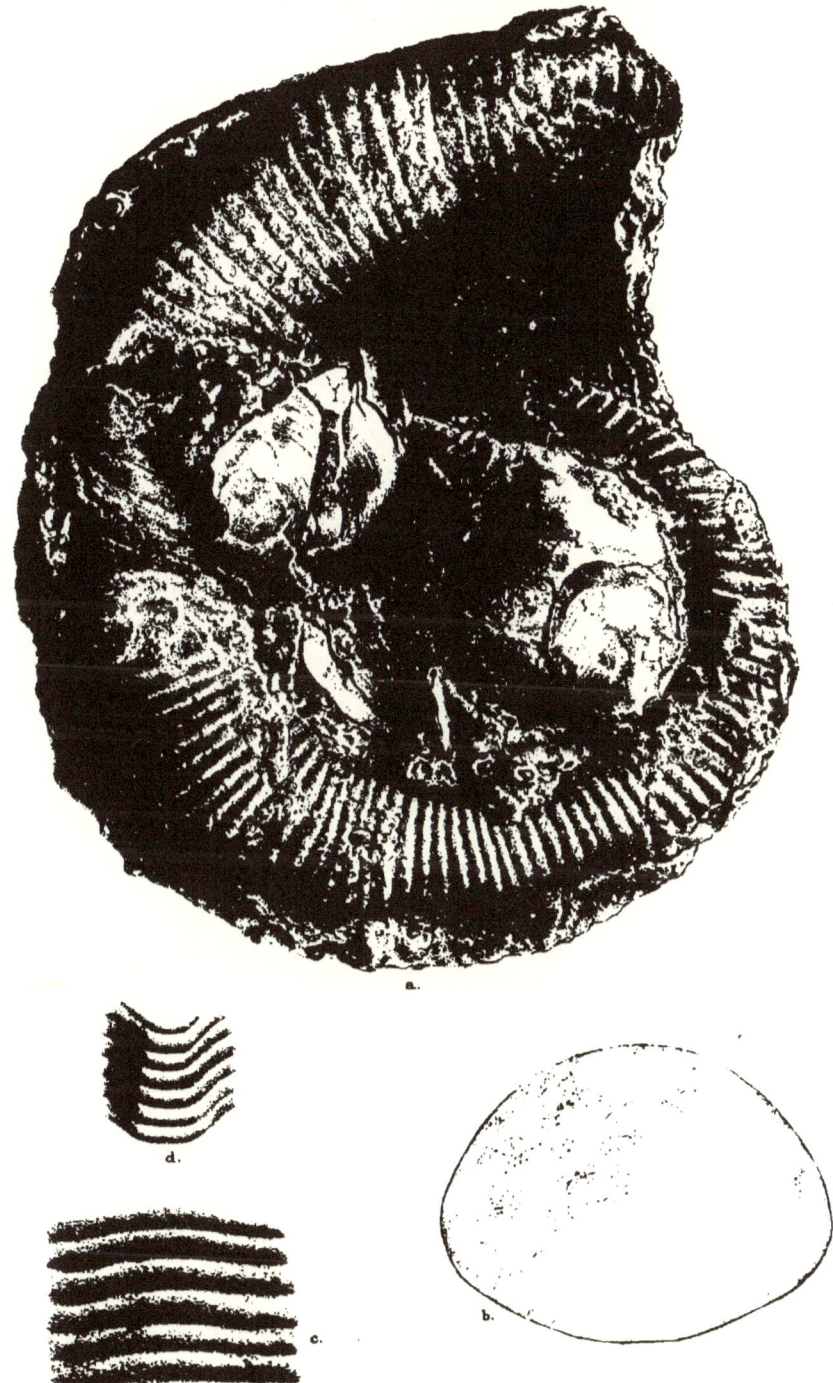

a.

d.

c.

b.

A S Foord del et lith.

Mintern Bros imp

PLATE III.

Pl. III.

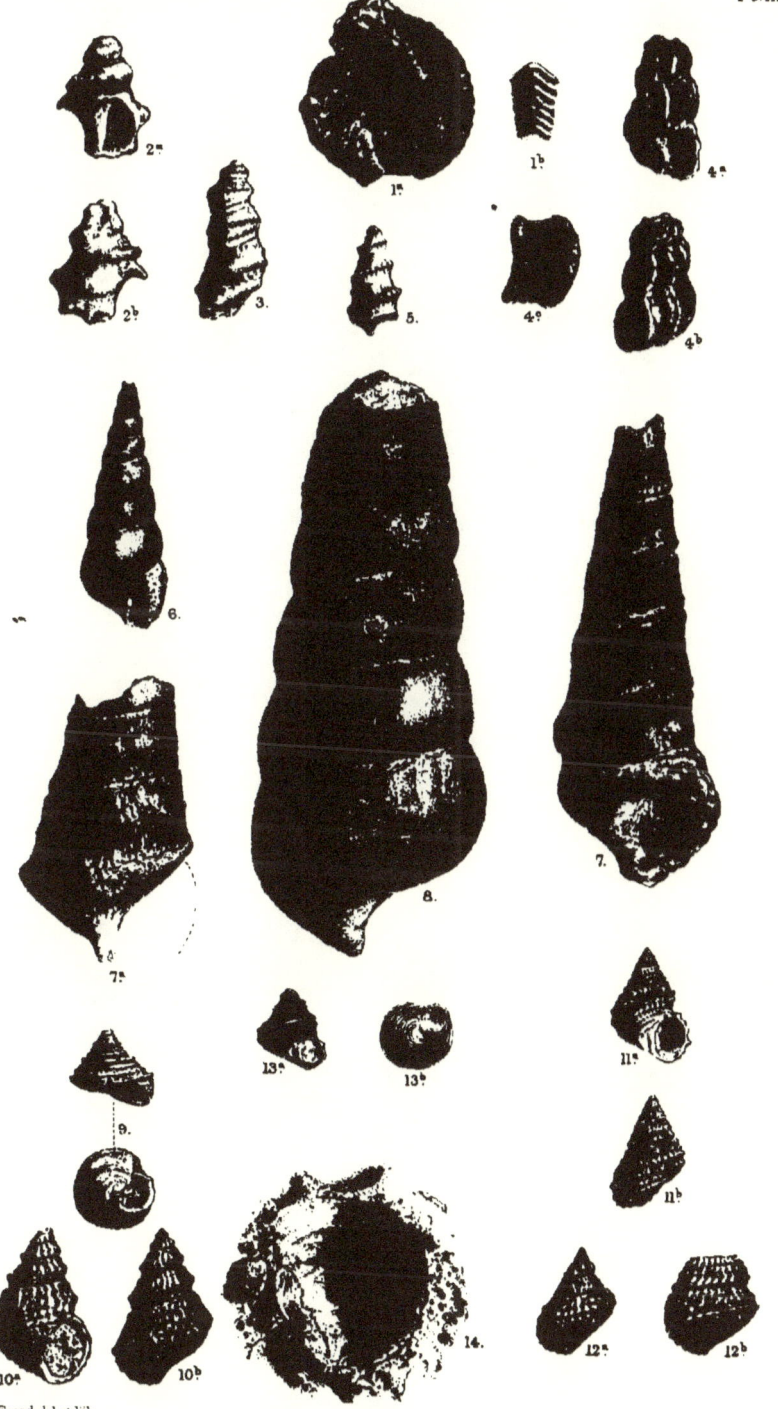

A.S.Foord del.et lith.

Mintern Bros. imp.

PLATE IV.

Pl. IV.

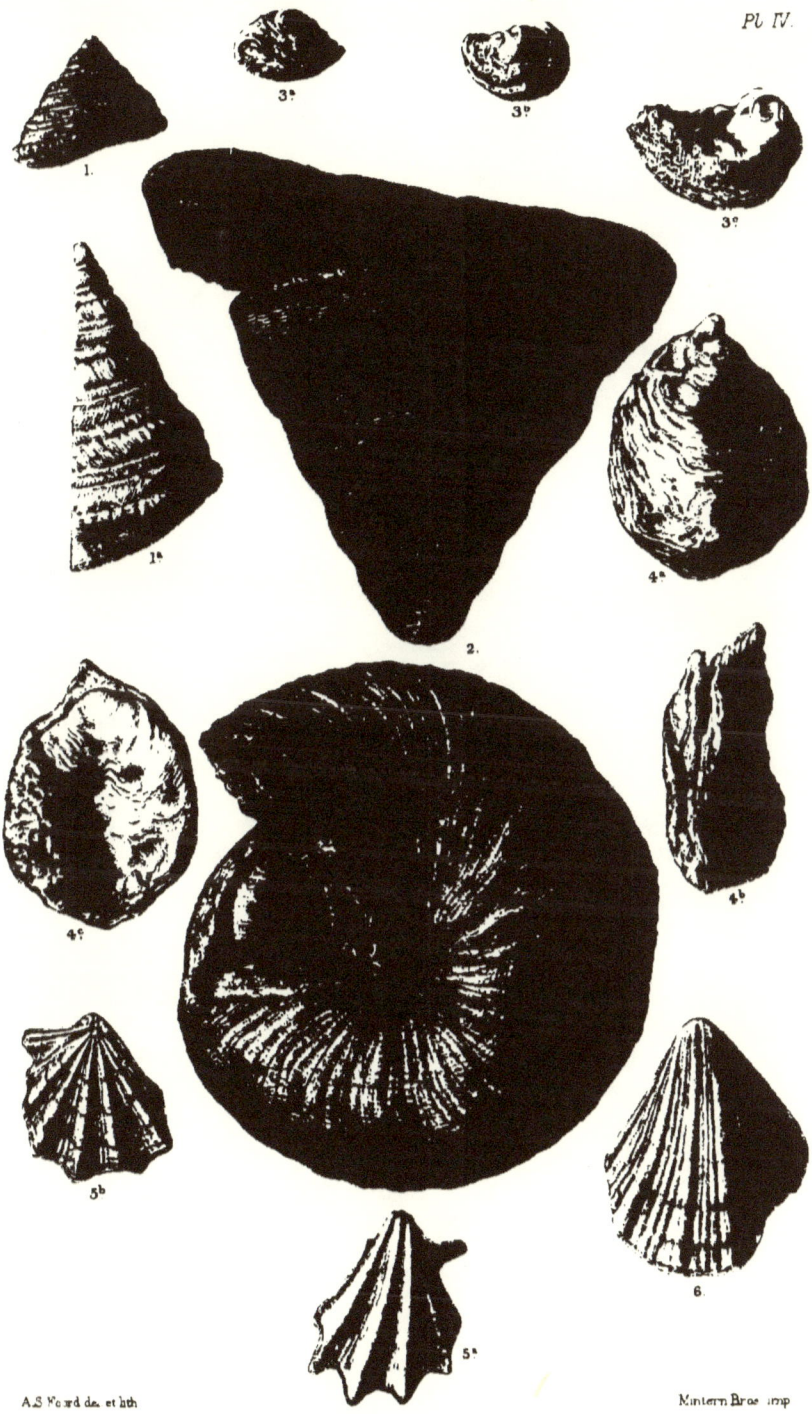

PLATE V.

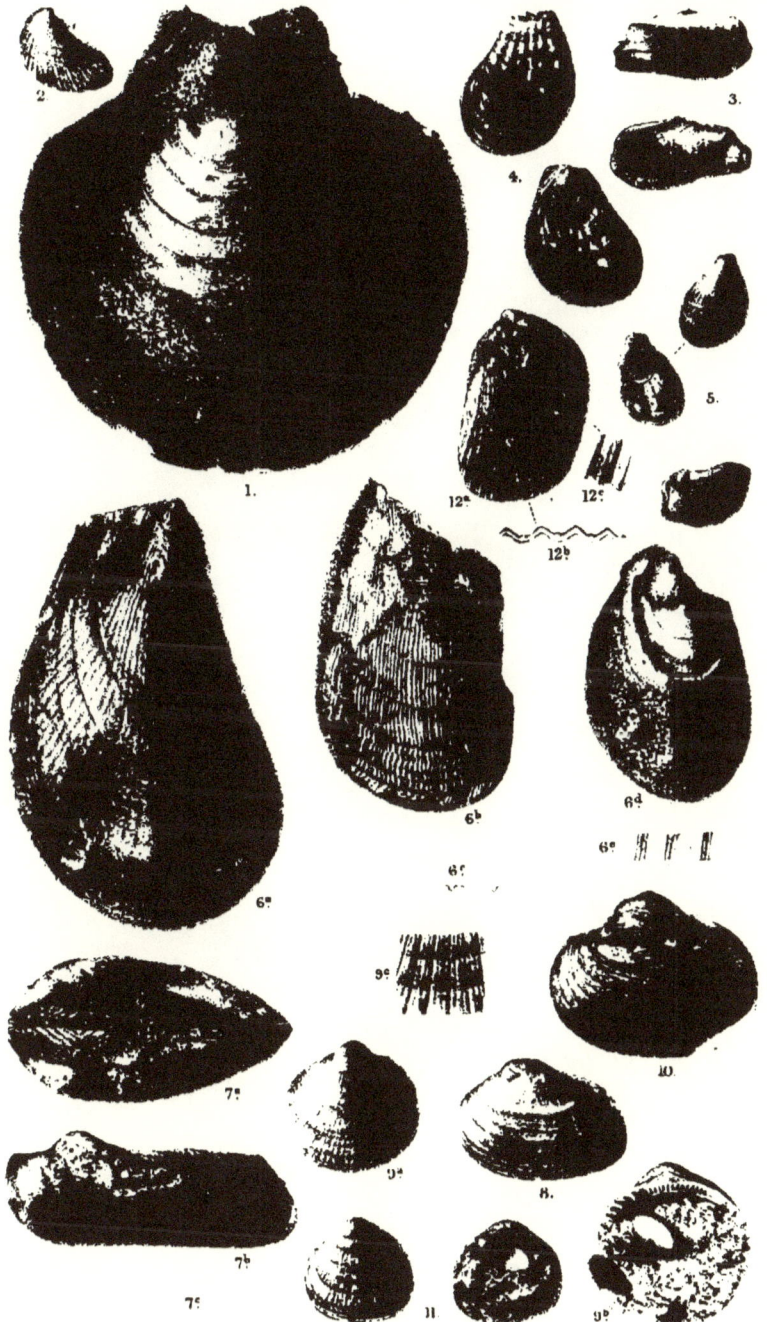

PLATE VI.

Pl. VI.

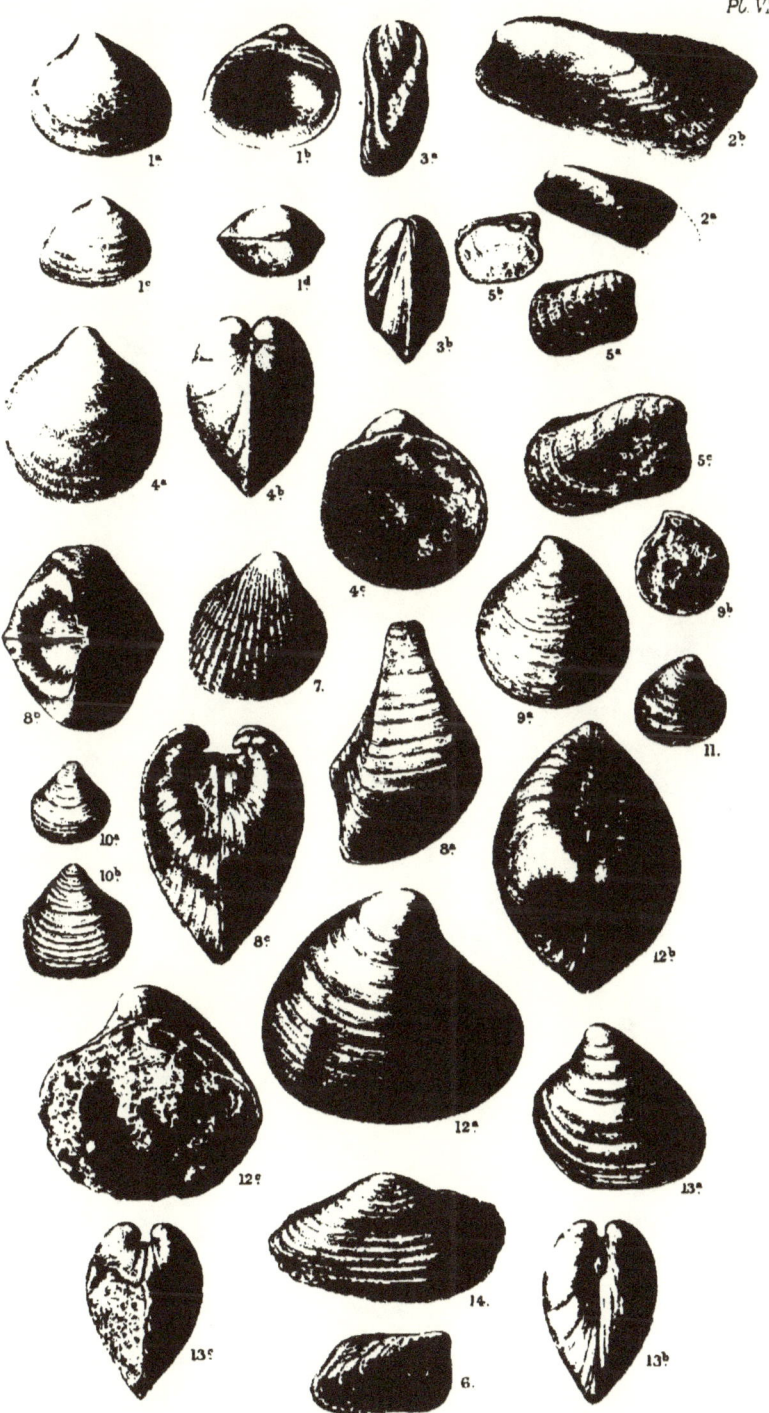

A S Foord del et lith.

Mintern Bros imp.

PLATE VII.

Pl. VII.

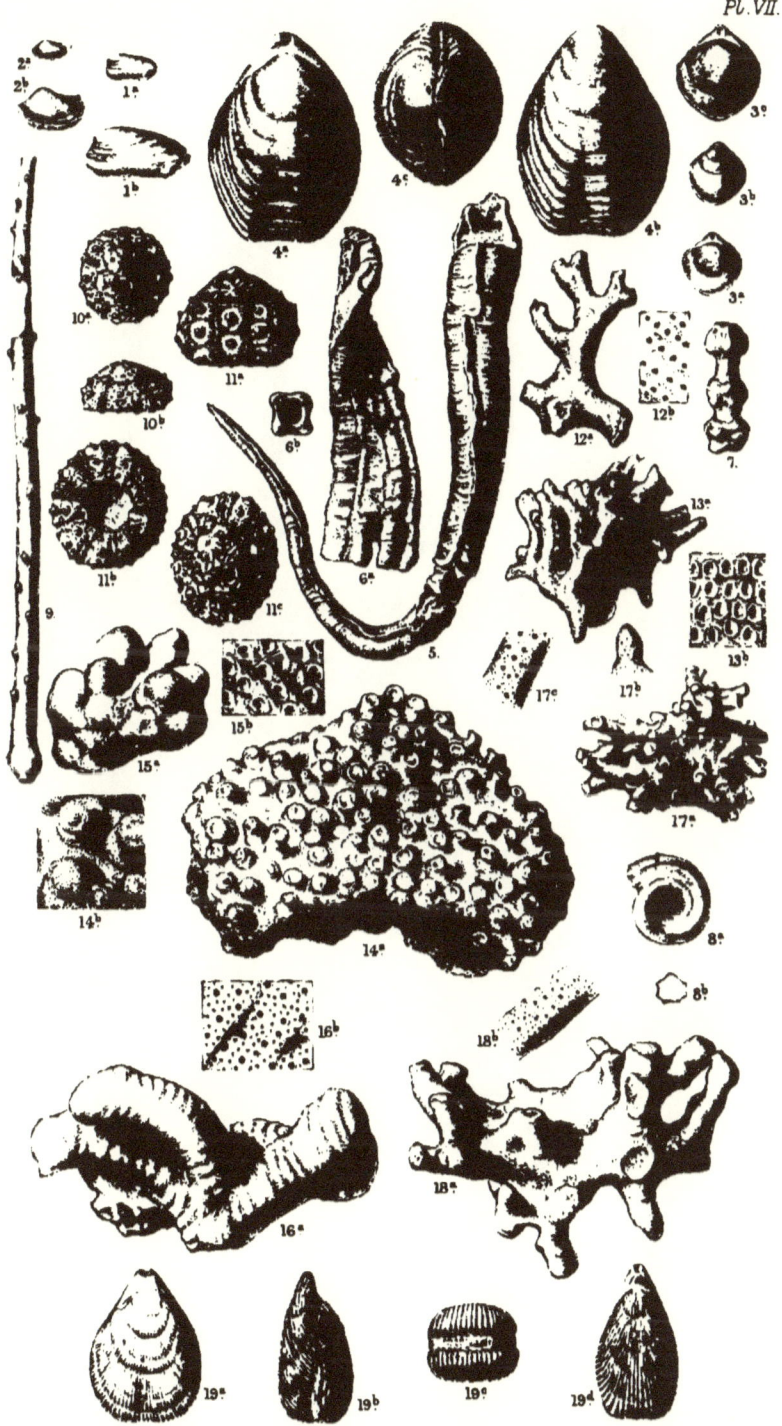

Mintern Bros imp.

PLATE VIII.

Fig.

1. *Plocoscyphea pertusa*, Geinitz, Brickhill. 1 *b*, structure of the same, enlarged.

2. *Verticellites annulatus*, Kpng., Upware.

3. *Verticellites clavatus*, Kpng., Upware.

4. *Corynella nodosa*, Kpng., Brickhill.

5. *Corynella*, sp., Brickhill.

6. *Pachytiloda*, sp., Upware.

7. A small fruit (?) ferruginised, Upware.

8. *Cucullaea vagans*, Kpng., Upware (derived). 8 *b*, the internal cast.

9. *Cucullaea Donningtonensis*, Kpng., Upware (derived).

10. *Terebratula ovoides*, Sby., var. *rex*, Ray Lankester, Upware. 10 *b*, side view.

11. *Ammonites biplex*, Sby., Upware (a phosphatised specimen, much eroded).

Pl. VIII.

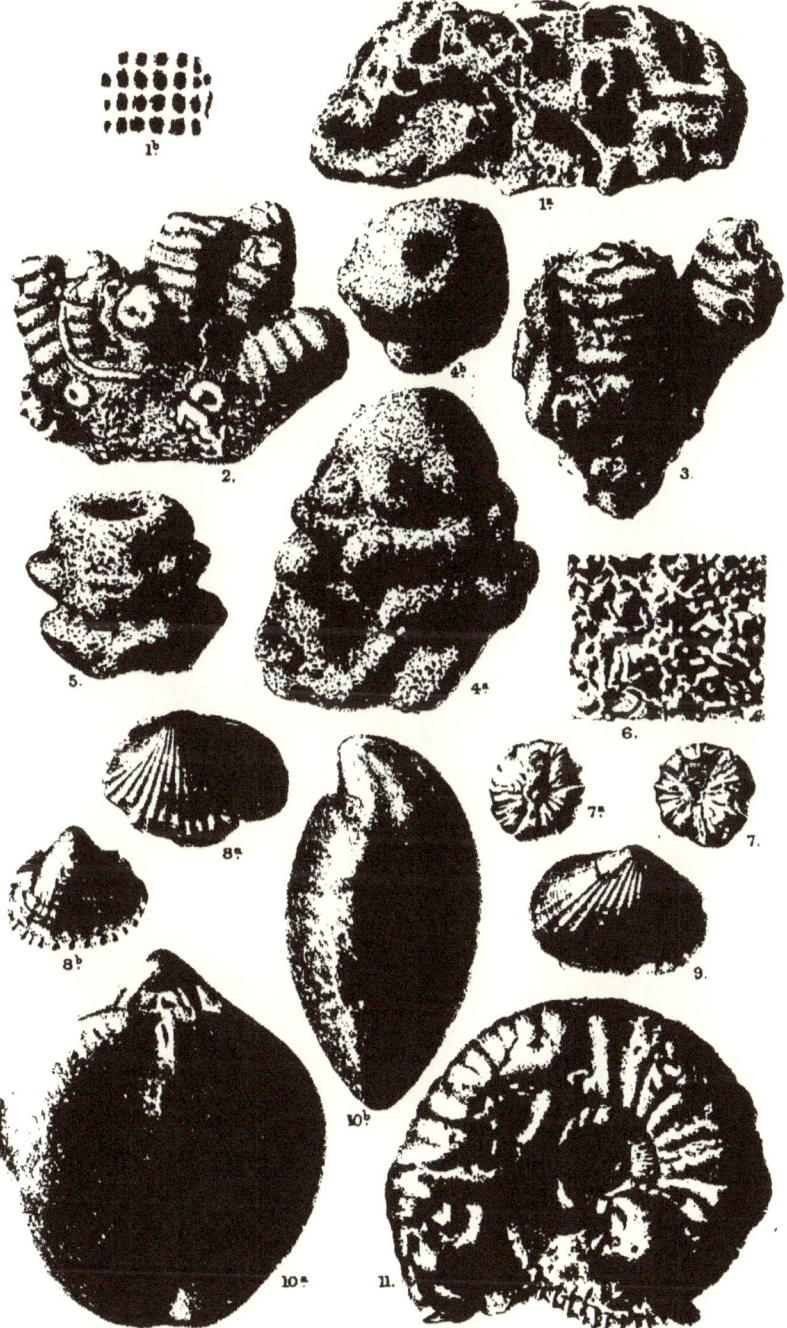

A.S.Foord del.et lith.

Mintern Bros imp.

www.ingramcontent.com/pod-product-compliance
Lightning Source LLC
Chambersburg PA
CBHW020616030726